The Promise of AI

A guidebook for the policymakers to navigate in the flat world.

Sanjiv Goyal

Adroit Capital

Copyright © 2023 by Sanjiv Goyal, All rights reserved. No part of this publication may be reproduced, distributed, or transmitted in any form or by any means, including photocopying, recording, or other electronic or mechanical methods, without the prior written permission of the publisher, except in the case of brief quotations embodied in critical reviews and certain other noncommercial uses permitted by copyright law. For permission requests, write to the author at sanjiv@adroit.cc, addressed "Attention: Permissions."

ISBN: 9798854456302

Dedication

To my beloved parents, cherished siblings, and my precious children, Yash and Tia. Your unwavering inspiration fuels my every day, and your presence infuses boundless motivation into my journey.

The Promise of AI

Contents

Introduction 9

Part 1: History of AI

Evolution of AI 17

Part 2: Our Future with AI

AI & future of the flat world 39

Understanding Citizens' Neurological Behavior Patterns 45

AI Driven Policymaking in The Flat world 49

The Global Imperative and Policy Cooperation now and Beyond 57

Pioneering a Collective Vision for a Thriving Future 63

Part 3: AI Ethics and Bias

Ethics and Unbiased Approaches 69

Data Ethics and Transparency 77

Bias Mitigation 83

Privacy and Data Protection Privacy of Citizens' Data 87

Human oversight and accountability	91
Inclusive AI Development	95
Continuous Auditing and Evaluation	99
Case Studies	103
Embracing the Ethical Imperative	115
Understanding Citizens Neurological Behavior Patterns	119

Part 4: The Promise of AI for governments and citizens

The Promise of AI for Government	125
The Challenges of Adopting AI in Government	129
Empowering Tomorrow's Guardians	135
AI for Healthcare	139
Empowering Lifelong Learning in a Transformative Era	143
Pioneering Smart Mobility in a Connected World	149
AI for Financial Services	155
Building a Sustainable Tomorrow	161

Crafting a Tailored AI Vision for Your Department	167
AI Framework for Policy Makers	171

Part 5: Envisioning Tomorrow's Possibilities

Art of possible	177
Mavericks' AI leaders	185
Resources	189
AI-Powered Tools for policymakers	193
Assess Your AI Skills	205

The Promise of AI

Preface

Welcome to the captivating world of AI Adventures! As the author of this book, I am thrilled to be your guide on this exciting journey into the future of Artificial Intelligence (AI) and its boundless applications. Together, we'll explore the marvelous possibilities AI holds for transforming industries, enriching lives, and shaping a more inclusive and sustainable tomorrow.

Picture a world where self-driving cars navigate bustling city streets, effortlessly zipping through traffic jams, and AI-powered virtual assistants anticipate your needs with an uncanny understanding of your preferences. Imagine personalized medical treatments tailored to your unique genetic makeup and AI-driven tutors helping you unlock the keys to endless knowledge.

In our AI Adventure, we'll witness how AI becomes an omnipresent companion, revolutionizing various sectors. We'll dive into finance, where robo-advisors like Wealth front and Betterment empower investors with personalized financial guidance, and PayPal's AI-powered fraud detection safeguards your hard-earned money.

Once confined to classrooms, education expands beyond boundaries as AI-driven platforms like Duolingo and Khan Academy provide individualized learning pathways, turning you into a lifelong learner. We'll explore how AI shapes how we move as autonomous vehicles hit the roads, making transportation safer, greener, and more efficient.

But our journey continues; we'll venture into public safety, where AI-powered analytics revolutionize crime prevention and emergency response. Additionally, we'll witness AI donning its environmental superhero cape, aiding in wildlife conservation, monitoring climate change, and promoting sustainable practices.

Throughout this book, you'll encounter an array of AI companions, from chatbots that entertain and assist you like IBM's Watson and Bank of America to AI-driven medical marvels like IBM Watson for Oncology, supporting doctors in diagnosing and treating complex diseases. But this adventure isn't just about the awe-inspiring applications of AI; we'll also grapple with its ethical dimensions. Together, we'll ponder how to ensure AI's responsible use, safeguarding our values and preserving human dignity.

Dear readers, this book is not a collection of technicalities; it's a journey into the future. I aim to make this fascinating world of AI accessible to all, whether you're an expert in the field or someone eager to peek into tomorrow's possibilities. Through clear explanations and relatable examples, we'll unravel the mysteries of AI and its transformative potential.

AI holds the promise of sculpting a future that transcends our current understanding. Its capabilities are as awe-inspiring as they are enigmatic, capable of reshaping industries, enriching lives, and reshuffling the deck of human potential. In this book, we're here to unravel these mysteries and delve into the incredible stories and discoveries that await us.

Together, we'll decode AI's intricate language, uncover its hidden treasures, and navigate its uncharted territories.

We'll confront the challenges and embrace the triumphs, knowing that AI is both an ally and a challenge. Its boundless capabilities, tempered with our ethical compass, offer the potential for unprecedented progress.

So, let us embark on this quest with open hearts and minds, ready to absorb AI's wonders and wisdom. As we peer into the future, remember that the magic happens in uncharted territories. Our journey begins now, and the possibilities are as limitless as our collective imagination. The future beckons, and with AI as our companion, there are no boundaries to what we can achieve or explore. Let's dive in!

Let the AI Adventures begin!

With excitement and curiosity,

Sanjiv Goyal

www.linkedin.com/in/sanjivgoyal
www.youtube.com/sanjivgoyal
www.sanjivgoyal.com

The Promise of AI

Introduction

Get ready to embark on an exhilarating journey through Artificial Intelligence (AI) in our action-packed book, where the future meets imagination, and the possibilities are endless! Join us as we dive into the extraordinary applications of AI, from soaring with autonomous drones to unraveling the mysteries of AI-powered chatbots.

This isn't your average AI textbook! We're ditching the jargon and technical complexities, and instead, I'll be your trusty AI tour guide, making this exciting world easy to comprehend and a lot of fun! Chapter by chapter, we'll venture into diverse sectors, exploring how AI is shaping industries, enriching lives, and even safeguarding the environment. You'll discover how AI superheroes are revolutionizing healthcare, education, finance, and public safety while preserving our beloved planet.

Ever wondered what it's like to have a personal AI tutor or an AI assistant? How about hopping aboard self-driving cars that navigate the roads with the finesse of a race car driver? Buckle up because we'll be journeying through a high-tech realm of possibilities where imagination and reality collide.

We'll dive into the power of AI's predictive analytics, unleashing its secrets to foresee climate changes, spot trends in financial markets, and even anticipate your next shopping spree! Don't worry; you won't need a crystal ball, just a dash of AI magic! And we promise that the AI adventure is only complete with meeting some friendly AI companions. You'll get up close and personal with chatbots that will make you smile, chatbot advisors that will be your financial gurus, and intelligent tutors that will guide you through the maze of knowledge. But it's not all robots and algorithms; we'll be exploring the ethical dimensions of AI, too. As we soar through AI's boundless horizons, we'll ponder the importance of responsible AI and how we can ensure that these AI-powered superheroes uphold our values and ethical standards.

So, whether you're an AI enthusiast, a curious explorer, or someone who wants to peek into the future, this book has something in store for you. We invite you to join us on this thrilling AI adventure, where the possibilities are endless, and the knowledge is as boundless as the universe itself. Are you ready to experience the Marvelous World of AI Adventures? Let's dive in, have a blast, and unlock the secrets of this extraordinary frontier together! Grab your AI cape and prepare to soar into the world of AI like never before! The future awaits, and with AI as our guide, it's bound to be nothing short of amazing! Let the adventure begin!

This book is divided into five parts.

1. History of AI

In the vast annals of human ingenuity, Artificial Intelligence (AI) emerges as a marvel of our ceaseless drive to replicate our own cognitive capacities in machines. Spanning from age-old myths of automatons to contemporary advanced technologies, AI's evolution is a testament to our enduring pursuit of understanding and emulating intelligence. From its embryonic origins in ancient tales of mechanized beings, AI matured in the mid-20th century, bolstered by luminaries like Alan Turing and his revolutionary "universal machine" concept. With the advent of computers, these foundational ideas materialized, catalyzing an avalanche of progress, from the pioneering efforts of figures like John McCarthy to the oscillating waves of optimism and skepticism, shaping AI's intriguing history.

2. Our future with AI

Advancements in neuroscience and AI are changing the way policymakers shape policies and enforce them. This article explores how these changes are revolutionizing public institutions. Additionally, global cooperation is crucial for a sustainable future beyond 2050, as emphasized by Thomas Friedman's "The World Is Flat" approach. The link between neuroscience and policymaking is complex, but case studies demonstrate the potential of using behavioral economics to guide national governance.

3. AI Ethics and Bias

Our exploration delves into the relationship between citizens' neurological behavior patterns and ethical AI policymaking. As AI increasingly affects our lives, we examine its impact on public safety, healthcare, education, transportation, financial services, and environmental protection. Real-world case studies reveal ethical dilemmas and biases, highlighting the importance of an ethical AI framework that empowers governments and institutions. We showcase the GDPR's data protection standard and emphasize the urgency of setting policies for a fair and just future.

4. The promise of AI for governments and citizens

AI can revolutionize governance for governments by facilitating evidence-based decision-making and improving public services. Its ability to harness vast amounts of data from diverse sources empowers policymakers to develop more targeted policies. AI technologies can offer personalized and efficient services, automate tasks, aid law enforcement agencies in proactive crime prevention, and foster a more inclusive education system. The benefits of AI for governments are vast.

5. Envisioning Tomorrow's Possibilities

AI can revolutionize governance for governments by facilitating evidence-based decision-making and improving public services. Its ability to harness vast amounts of data from diverse sources empowers policymakers to develop more targeted policies. AI technologies can offer personalized and efficient services, automate tasks, aid law enforcement agencies in proactive crime prevention, and foster a more inclusive education system. The benefits of AI for governments are vast.

The Promise of AI

Part 1

History of AI

Evolution of AI

In the vast landscape of human innovation, few frontiers have captivated our imagination and curiosity quite like Artificial Intelligence (AI). From ancient myths of automatons to the awe-inspiring technologies of today, the history of AI is a chronicle of humanity's unrelenting quest to create machines that emulate human intelligence. As we embark on a journey through time, we'll unravel the tapestry of achievements, setbacks, and monumental leaps that have shaped the evolution of AI. This journey is not just a tale of technological prowess; it's a testament to our insatiable desire to understand the essence of cognition and replicate it in our creations.

The origins of AI can be traced back to antiquity when ancient civilizations spun tales of mechanical beings that could perform tasks akin to human abilities. However, in the mid-20th century, AI truly began to emerge as a distinct field of scientific inquiry. Visionaries like Alan Turing laid the foundation with his groundbreaking concept of a "universal machine" that could simulate any human cognitive process. The birth of computers brought these ideas to life, sparking a cascade of innovations that would define the AI landscape.

From the early days of simple algorithms to today's complex neural networks, AI has undergone a metamorphosis that defies expectations. We'll traverse the terrain of early AI pioneers like John McCarthy, who coined the term "Artificial Intelligence" and organized the Dartmouth Conference in 1956. This pivotal event marked the official birth of AI as an academic discipline. We'll delve into the optimism of the 1960s when researchers believed that human-level AI was just around the corner, only to be disillusioned during the "AI winter" of the 1970s and 80s.

But the story doesn't end there. In the late 20th century, we witnessed the resurgence of AI through expert systems, symbolic reasoning, and machine learning. As we entered the 21st century, the fusion of big data, powerful computing, and advanced algorithms breathed new life into AI, leading to unprecedented breakthroughs in natural language processing, image recognition, and robotics. Today, AI permeates our lives in ways unimaginable, from digital assistants that respond to our voices to self-driving cars navigating our streets.

Yet, as we marvel at AI's modern marvels, we must also confront ethical dilemmas, biases, and questions about the role of AI in our society. This journey through the history of AI is not just a recollection of technological advancements; it's an exploration of the human spirit's ceaseless pursuit of understanding, innovation, and the quest to unlock the enigma of intelligence itself. As we embark on this journey, let us unravel the pages of AI history with a sense of wonder, humility, and anticipation of the future.

Let's deep dive into its early beginnings to the present day. Let me cover significant milestones in AI research and the challenges and successes encountered.

1.1 Introduction to Artificial Intelligence - Unveiling the Power of Machine Intelligence

Artificial Intelligence (AI), the term that conjures images of machines simulating human-like intelligence, has evolved from a science fiction concept to a transformative force underpinning our digital age. As we embark on a journey through the annals of AI history, we're invited to witness the remarkable evolution of machines capable of learning, reasoning, and problem-solving.

Imagine the awe-inspiring moment when IBM's Watson, in 2011, triumphed over human champions in the American quiz show "Jeopardy!" Watson's ability to understand natural language, analyze vast volumes of data, and generate accurate answers in seconds marked a pivotal step in AI's evolution. However, this victory was merely the culmination of decades of innovation and experimentation.

Delve into the 20th century, and you'll encounter the visionary Alan Turing, whose conceptualization of the Turing Test in the 1950s laid the foundation for assessing a machine's ability to exhibit intelligent behavior indistinguishable from that of a human. Turing's work set the stage for AI's birth and ignited discussions about the essence of human intelligence.

Fast forward to the 1956 Dartmouth Workshop, where luminaries like John McCarthy set out to create a new field of study - artificial intelligence. This was the genesis of a journey that embraced symbolic AI, where researchers aimed to replicate human thought processes using logical rules and symbols. Programs like Logic Theorist and General Problem Solver emerged as AI's pioneers during this era, proving machines could indeed think logically.

In the 1960s, Joseph Weizenbaum's ELIZA introduced the concept of natural language processing, a precursor to modern conversational agents. Then came the AI winters, periods of reduced funding, and optimism due to technology limitations. However, these challenges ignited a metamorphosis in AI research.

The late 20th century saw a resurgence with the rise of machine learning, particularly neural networks, and evolutionary algorithms. These approaches paved the way for advancements in pattern recognition and data analysis. A watershed moment arrived in 1997 when IBM's Deep Blue defeated world chess champion Garry Kasparov, demonstrating that machines could outperform humans in complex strategic tasks.

Fast forward again to the 21st century, where deep learning, inspired by the human brain's neural structure, has unlocked unprecedented capabilities. Google's AlphaGo stunned the world by defeating a world Go champion, showcasing the immense potential of AI's pattern recognition and decision-making prowess.

Throughout AI's history, we've witnessed moments of awe and moments of reflection. We've pondered ethical concerns, debated the potential of artificial general intelligence, and embraced AI's potential to revolutionize industries. From expert systems to autonomous vehicles, AI's impact is tangible in healthcare, finance, education, and more.

As we explore the AI journey spanning the last 70 years, we are beckoned to acknowledge that AI has transcended fiction and become an integral part of our reality. Today, AI is shaping our present and illuminating a future where machines mimic human intelligence and elevate it, enhancing our lives and sparking new frontiers of discovery.

1.2 Early Beginnings - Pre-20th Century, Pioneering Automata and Ancient Insights

Before the digital age, the seeds of Artificial Intelligence were sown in the fascinating realm of ancient civilizations and mechanical marvels. Long before the advent of modern computers, ingenious inventors and thinkers laid the groundwork for machines capable of intelligent actions.

Imagine stepping into the ancient world of Alexandria, where the renowned mathematician Hero of Alexandria crafted intricate automatons - mechanical devices that moved and performed tasks using clever mechanisms. Hero's creations included mechanical birds that sang and statues that poured water, astounding people with their seemingly mystical abilities. These early endeavors into crafting automata were, in essence, the seeds of AI.

Fast forward to the Islamic Golden Age, where polymaths like Al-Jazari designed astonishing mechanical devices in the 12th century. Al-Jazari's "Book of Knowledge of Ingenious Mechanical Devices" featured a host of wondrous machines, from mechanical musicians to a humanoid "drinking servant." These inventions offered a glimpse into the fusion of human ingenuity and mechanized actions, foreshadowing the AI pursuits of the future.

While these ancient marvels showcased glimpses of machines mimicking human-like actions, it wasn't until the 17th century that René Descartes' musings on animal behavior hinted at the potential for mechanized reasoning. Descartes suggested that animals might be understood as intricate machines, marking a shift in how people contemplated the boundaries between intelligence and machinery.

Yet, the true catalyst for AI's early beginnings emerged during the 19th century's Industrial Revolution. Charles Babbage, an English polymath, conceptualized the Analytical Engine - a mechanical device with a punch card system for calculations. While never fully realized in his lifetime, Babbage's ideas laid the groundwork for programmable machines that could manipulate symbols and execute instructions, akin to the foundational principles of AI.

As we peer into these early chapters of AI's history, it's evident that the pursuit of machines with intelligence-like attributes has ancient roots. From ancient automatons that delighted onlookers to visionary inventors like Babbage, who envisioned the mechanization of logical tasks, the seeds of AI were sown long before silicon circuits and algorithms. These historical glimpses are a testament to human curiosity, imagination, and the persistent drive to bridge the gap between the mechanical and the intelligent.

1.3 Emergence of Computing and WWII Influence - Alan Turing's Genius and the Seeds of AI

The wheels of progress turned swiftly in the 20th century, especially during World War II, as the world grappled with unprecedented challenges. Amid the turmoil, a brilliant mind emerged that would lay the foundation for the field of Artificial Intelligence (AI) as we know it today. Enter Alan Turing, a mathematician, logician, and visionary whose contributions still resonate in our digital age.

Imagine a time when nations were embroiled in a global conflict. Alan Turing's genius was instrumental in deciphering the German Enigma code and sowing the seeds of AI. His groundbreaking work at Bletchley Park, where he devised the Turing Machine - a theoretical device capable of solving any computational problem through a series of logical steps - heralded the dawn of the computing age.

Turing's legacy was more than just codebreaking; it was a revolutionary perspective that machines could simulate intelligent human thought processes. His 1950 paper, "Computing Machinery and Intelligence," presented the now-famous Turing Test - a benchmark to determine a machine's ability to exhibit human-like intelligence. This pivotal concept ignited discussions that continue to shape AI's trajectory.

The Turing Test spurred innovation, leading to the creation of the first electronic digital computers. Turing's principles set the stage for early computing pioneers like John von Neumann, whose work on stored-program computers laid the foundation for programmable machines. These developments breathed life into the concept of machines that could not only crunch numbers but also execute logical operations - a cornerstone of AI's future.

Fast forward to the present, and we see Turing's influence resonating in every facet of our lives. Think of your smartphone, which operates on the same principles Turing envisioned decades ago - processing information, learning patterns, and even engaging in conversations through AI-powered virtual assistants. It's astonishing to realize that Turing's work, initially conceived during the chaos of war, has become an integral part of our daily existence.

As we journey through history, let's acknowledge Alan Turing's indelible mark on the development of AI. His brilliance transformed the computing landscape, laying the groundwork for machines that compute and simulate aspects of human intelligence. Turing's legacy is an inspiration, reminding us that innovation can flourish even amidst the most challenging circumstances and that his legacy continues to fuel AI's boundless potential in our interconnected world.

1.4 Dartmouth Workshop and Birth of AI (1956) - Igniting the Spark of Artificial Intelligence

The year was 1956, and a pioneering gathering was about to take place to shape the course of history. Imagine a time when computers were clunky, room-sized machines, and the notion of machines thinking like humans seemed like a far-off dream. Against this backdrop, the Dartmouth Workshop unfolded, birthing the field of Artificial Intelligence (AI) as we know it.

Picture a group of visionary thinkers converging at Dartmouth College, including John McCarthy, Marvin Minsky, Nathaniel Rochester, and Claude Shannon. Their mission was audacious yet exhilarating: to explore the idea of creating machines that could simulate human intelligence. The term "Artificial Intelligence" was coined, and thus began a journey of exploration into the realm of machine minds.

The Dartmouth Workshop marked the pivotal moment when AI transitioned from a lofty philosophical concept to a field with a clear direction. McCarthy's vision of "making a machine behave in ways that would be called intelligent if a human were so behaving" laid the groundwork for what would come.

One of the earliest fruits of the Dartmouth Workshop was the Logic Theorist, an AI program that could prove mathematical theorems. This creation demonstrated that machines could engage in problem-solving tasks requiring human-like reasoning. The success of Logic Theorist emboldened researchers to delve deeper into developing AI systems that could perform tasks that once seemed exclusive to human cognition.

Fast forward to today, and the legacy of the Dartmouth Workshop is everywhere. Whenever you use voice recognition technology like Siri or navigate through recommendation systems like those on Netflix or Amazon, you're experiencing the seeds planted at that pivotal event.

The chess-playing AI Deep Blue's victory over grandmaster Garry Kasparov in 1997, the inception of virtual assistants like Google Assistant, and the rise of autonomous vehicles - all these marvels trace their lineage back to that historic gathering.

The Dartmouth Workshop was not just a meeting but the birthplace of a movement. It was a moment when a group of brilliant minds dared to dream of machines that could think, learn, and solve problems like humans. It set the course for AI's trajectory, sparking a journey of innovation, challenges, and breakthroughs that continue to shape our world today. It reminds us that audacious dreams have the power to reshape reality, and the legacy of those visionary thinkers is alive in the very fabric of our digital age.

1.5 The AI Summers and Winters: Triumphs, Setbacks, and Resilience

In the narrative of Artificial Intelligence (AI), triumphs and setbacks have intertwined, propelling the field forward in a dance of innovation and reflection. Imagine a roller coaster ride through the 20th century, where the promise of AI soared to dizzying heights during "AI Summers" and faced cold, challenging periods during "AI Winters."

During the late 1950s and 1960s, AI was a young and ambitious field. Researchers embraced symbolic AI, hoping to replicate human intelligence through logical rules and representations. This was the era of "AI Summer," characterized by a surge in optimism and investment. It birthed systems like ELIZA, a chatbot that simulated human conversation, and DENDRAL, which could deduce the structure of organic molecules - feats that seemed miraculous at the time.

The 1960s and 1970s, however, ushered in an "AI Winter." The initial enthusiasm collided with the challenges of complex problem-solving and the realization that AI's grand goals were far from reach. Funding dwindled, enthusiasm waned, and AI seemed to lose its luster. But the resilience of AI researchers and their unwavering belief in the potential of their field kept the spark alive.

Fast forward to the 1980s and 1990s, a period marked by a resurgence of AI, another "AI Summer." Expert systems became the rage, as AI systems designed to emulate human expertise were employed in medical diagnosis, finance, and other domains. Pioneering achievements like IBM's Deep Blue defeating chess champion Garry Kasparov in 1997 further propelled AI's reputation.

The new millennium saw machine learning and neural networks blossoming, setting the stage for the current AI revolution. AI systems like IBM's Watson showcased the potential of machine learning in language understanding, enabling the computer to beat human champions in the quiz show "Jeopardy!"

These triumphs, however, were not without their challenges. The mid-2010s marked another "AI Winter" as hype exceeded reality. The over-promise of AI capabilities led to skepticism and a more measured approach to implementation.

But here's the remarkable part: every AI winter has been followed by a spring. The lessons learned during periods of reflection have fueled new innovations and breakthroughs. The AI of today is far from the AI of the past. Autonomous vehicles roam our streets, AI-driven medical diagnostics revolutionize healthcare, and voice assistants make our lives more convenient.

So, as we traverse through AI's history, let's recognize that triumphs and setbacks are the fabric of progress. The dance between AI Summers and Winters has shaped the resilience and wisdom of the field. It reminds us that the innovation journey is not a straight line but a dynamic ebb and flow. Each challenge overcome, each setback turned into an opportunity, has brought us to the precipice of AI's boundless potential. The past's lessons are the guideposts for an AI-powered future, where challenges inspire growth, and triumphs fuel our journey towards greater horizons.

1.6 AI's Recent Resurgence: The Age of Deep Learning and Beyond

Imagine a world where machines learn from data where computers can understand images, language, and even art. This is the realm of AI's recent resurgence, driven by the remarkable progress of deep learning. Fasten your seatbelts as we journey through the awe-inspiring landscape of modern Artificial Intelligence (AI).

The turning point came in the early 2010s when researchers harnessed the power of neural networks, mirroring how the human brain processes information. Deep learning, a subset of machine learning, emerged as the driving force behind AI's recent resurgence. The breakthroughs were dazzling: machines that could recognize faces, understand natural language and even play complex games like Go.

Take the case of ImageNet, an image classification challenge. In 2012, the deep learning model AlexNet astounded the AI community by reducing the error rate significantly, surpassing human accuracy. This marked a seismic shift, proving that neural networks could learn complex patterns and outperform traditional algorithms in tasks that seemed uniquely human.

Then came the jaw-dropping moment of 2016. AlphaGo, developed by DeepMind, a subsidiary of Google, defeated the world champion Go player Lee Sedol. Go, an ancient board game with more potential positions than atoms in the observable universe was considered an ultimate test of human intuition. AlphaGo's victory showcased the immense potential of deep learning to crack complex, non-linear challenges.

Fast forward to today, and the AI landscape continues to evolve. Conversational AI systems like GPT-3 generate human-like text, raising the curtain on AI-generated content from creative writing to programming assistance. Self-driving cars navigate city streets powered by intricate neural networks, deciphering their surroundings in real-time.

This resurgence, however, is more than just technological marvels. It's a testament to human ingenuity, fueled by collaborative research, data-driven insights, and an unquenchable thirst for pushing boundaries. It's about transforming the impossible into reality and reshaping industries.

As we stand on the cusp of the AI revolution, we must remember that every breakthrough is an accumulation of small steps of countless hours spent tinkering with algorithms and fine-tuning models. The narrative of AI's recent resurgence speaks to humanity's capacity to learn, adapt, and innovate. It's not just about machines mimicking human-like behavior; it's about pushing the frontiers of knowledge and transforming how we interact with technology.

So, imagine a world where machines understand, learn and create. It's not science fiction; it's the dazzling reality of our AI-driven present and the promise of an even more astonishing AI-powered future. The journey has been exhilarating, and the path ahead is illuminated by the collective brilliance of researchers, innovators, and dreamers who continue to redefine the boundaries of possibility.

1.7 AI's Ubiquitous Impact: From Siri to Self-Driving Cars

Imagine a world where your smartphone understands your voice commands, where cars navigate themselves through busy streets, and where recommendations for movies and products seem eerily attuned to your preferences. This is the realm of AI's ubiquitous impact, where technologies we use daily are powered by artificial intelligence, often without us even realizing it.

Let's start with Siri, your virtual assistant on Apple devices. When you ask Siri questions, it listens, processes your words, and provides relevant information or executes tasks. This seemingly magical interaction is a product of

AI's natural language processing capabilities. Siri's ability to understand spoken language and respond intelligently relies on sophisticated algorithms that decode speech patterns, identify keywords, and generate coherent responses.

Now, consider the remarkable progress in self-driving cars. These vehicles, equipped with a plethora of sensors and cameras, can navigate through complex traffic scenarios, make split-second decisions, and even park themselves. AI algorithms process real-time data from these sensors, recognizing pedestrians, other vehicles, and road signs. This allows the car's "brain" to make informed decisions, like slowing down when a pedestrian steps onto the road or changing lanes to avoid an obstacle.

AI also permeates our online experiences. When you browse Amazon or Netflix, have you noticed how the recommendations seem tailor-made for your tastes? This is AI at work, analyzing your previous choices and preferences to suggest products or content you're likely to enjoy. These recommendation systems use complex algorithms to uncover patterns in your behavior, refining their suggestions with each interaction.

The financial sector has also embraced AI for fraud detection. Imagine the massive amounts of data banks and credit card companies handle daily. AI algorithms sift through this data, identifying unusual patterns that might indicate fraudulent transactions. If you've ever received a call from your bank asking if a particular transaction was authorized, that's AI working behind the scenes to protect your financial interests.

The common thread in these examples is AI's ability to process enormous amounts of data and make informed decisions or predictions. It's about enhancing efficiency, accuracy, and convenience in our daily lives. AI's ubiquity isn't just confined to tech giants; it's woven into the fabric of countless industries, from healthcare to finance to entertainment.

As we navigate the landscape of AI's impact, it's crucial to understand that this technology isn't isolated from us; it's a part of our lives. It's not about replacing human capabilities; it's about augmenting them making our interactions with technology more intuitive, seamless, and efficient. The AI-powered future is already here, enhancing how we live, work, and play and inviting us to explore endless possibilities.

1.9 The Road Ahead: Charting the Future of AI; Imagine a future where AI-powered robots coexist with humans, smart cities optimize resources, and industries harness AI's potential to solve complex global challenges. This section takes you on a journey into the exciting possibilities of AI's future. It explores how it can reshape industries, drive innovation, and even redefine what it means to be human.

Picture this: You're walking through a city where traffic lights adjust in real time based on traffic patterns. Energy consumption is optimized. Waste is managed efficiently.

This is the vision of a smart city, where AI orchestrates urban life to make it more sustainable and livable. Singapore's "Smart Nation" initiative is a real-world example, using AI and data analytics to transform the city-state into a technologically advanced urban hub.

Industries across the spectrum are also embracing AI to drive innovation. Take agriculture, for instance. AI-powered drones can survey crops, identify areas needing attention, and optimize irrigation and pesticide use. This not only increases crop yield but also reduces environmental impact. In the financial sector, AI algorithms analyze market trends to make real-time investment decisions, potentially revolutionizing how we manage our finances.

The field of healthcare is poised for a significant transformation. Imagine a doctor receiving real-time insights about a patient's health based on wearable devices and continuous monitoring. AI can analyze this data, alerting medical professionals to anomalies and potential health risks. This proactive approach to healthcare has the potential to save lives and improve the quality of care.

The rise of AI also intersects with the future of work. While AI technologies can automate routine tasks, they also open new avenues for human creativity and problem-solving. Imagine collaborating with AI systems to develop innovative solutions where machines handle repetitive tasks while humans focus on strategic thinking. This synergy between humans and AI can create a more dynamic and fulfilling work environment.

However, as we chart this AI-powered future, navigating potential challenges is essential. Ensuring that AI systems are secure, unbiased, and aligned with human values is crucial. Ethical considerations, privacy safeguards, and regulations will be pivotal in shaping this future.

The road ahead is paved with opportunities and challenges. It's a journey of exploration, innovation, and adaptation. The AI-powered future isn't a distant dream; it's a reality unfolding before our eyes. By embracing AI's potential, fostering collaboration, and making informed decisions, we can collectively steer this technological evolution toward a future that benefits humanity.

Conclusion

The history of Artificial Intelligence is a narrative of audacious dreams and remarkable achievements, punctuated by challenges that pushed the boundaries of human ingenuity. Our expedition through the annals of time reveals an unquenchable thirst for understanding the intricacies of intelligence, leading to the birth of an entire field that has forever altered our relationship with technology.

From its nascent origins, where myth met machinery, to the resounding echoes of Turing's universal machine, AI emerged as a beacon of human potential. The pioneers who dared to dream of machines that think and learn like humans paved the way for an evolution that remains as vibrant as it is transformative.

The ebbs and flows of AI's journey—where enthusiasm met skepticism—underscore the complexity of the challenge. The AI winters were a testament to the tenacity required to bridge the chasm between aspiration and realization. Yet, these periods of reflection and recalibration led to breakthroughs that are now an inseparable part of our daily lives.

As we stand at the threshold of a new era, AI has woven itself into the very fabric of our existence. From self-driving cars navigating bustling streets to virtual assistants anticipating our needs, AI is both a marvel and a reminder of our boundless potential. But this journey is not without its ethical quandaries. The mirroring of human biases, the ethical considerations in AI warfare, and the need to ensure that AI serves humanity rather than controls it are challenges that require our utmost attention.

Reflecting on the history of AI, we are called to be stewards of innovation, custodians of ethics, and champions of responsible progress. The tale of AI is a testament to the indomitable human spirit that propels us forward, even when faced with the unknown. As we look to the future, let us heed the lessons of the past and the promise of the present to guide us in shaping a world where AI's potential is harnessed for the betterment of humanity.

As we close this history chapter, let us carry with us the legacy of the pioneers, the lessons of the AI winters, and the aspirations of a world transformed. The history of AI is not just a narrative; it's an invitation to imagine, innovate, and continue our shared odyssey of discovery in the ever-expansive realm of human intelligence and technological advancement.

The Promise of AI

Part 2

Our future with AI

The Promise of AI

AI & future of the flat world

How AI-Driven Policymaking and Global Cooperation is Shaping Our present and the Future Beyond 2050.

The 21st century has brought about a new era of governance with the convergence of neuroscience, AI, and policymaking. The behavior patterns of citizens' brains are now crucial in shaping policies, while AI is transforming decision-making and enforcement. Let's explore how these advancements revolutionize policymaking and enforcement for public institutions. With Thomas Friedman's "The World Is Flat" approach in mind, we'll also highlight the importance of global cooperation in creating a sustainable and prosperous future beyond 2050.

We are embarking on an exciting journey to explore AI-driven policymaking and neuro-politics. Let's uncover the fascinating link between neuroscience and policymaking. We'll dive deep into the complexities of cognitive biases and behavioral economics and learn how our neurological intricacies shape our policy preferences and decision-making processes. With real-life case studies that demonstrate the impact of nudging health behavior, retirement savings policies, and using behavioral economics in policy design, we'll reveal the tremendous potential of neuroscience in guiding national governance.

I want to present the possibility with hard facts of AI-driven policymaking. Where data transforms into policy insights, transcending traditional governance paradigms. We witness the prowess of AI in action as it optimizes urban planning in Singapore's Smart Nation Initiative, empowers global citizens through Estonia's e-Residency program, enhances education policies through AI assessments, and transforms environmental policy through the United Nations AI for Earth Initiative. As the world becomes increasingly interconnected, we unveil the essence of Thomas Friedman's "The World Is Flat" approach, emphasizing how global policy cooperation becomes critical for us to transcend beyond the horizon of 2050.

The use of artificial intelligence in enforcing laws and regulations is becoming more prevalent. It is crucial to balance technological advancement with ethical considerations to harness the potential of AI responsibly. Collective wisdom and foresight are essential in achieving this balance.

AI has a crucial role in shaping global cooperation and diplomacy. It offers insights that steer foreign policies and drive multilateralism, such as the Paris Agreement's climate policies. We can witness transformative impacts on the world through international collaboration in AI governance. Examples include the OECD's AI Principles and the World Economic Forum's Global AI Council, representing a shared vision of responsible AI adoption.

As authoritative explorers of this brave new world, we embrace the Neurological Revolution and the power of AI, transcending borders and connecting nations with a shared mission: to forge a sustainable and equitable future beyond 2050. The Neurological Revolution unravels the essence of human behavior, and AI becomes the instrument through which enlightened governance shapes our collective destiny. Embrace this voyage, for together; we chart a course into a world where neuroscience and AI-driven governance are the heralds of a better tomorrow.

I have divided my understanding into four key aspects.

Understanding Citizens' Neurological Behavior Patterns

1. The Neuro politics Nexus: Unraveling the Connection Between Neuroscience and Policymaking.
2. Cognitive Biases and Decision-Making: How Our Brains Influence Policy Preferences
3. Leveraging Behavioral Economics: Nudging Policies to Align with Human Behavior
4. Neurotechnology and Public Policy: The Ethical Implications of Brain-Computer Interfaces

AI-Driven Policymaking: The World Is Flat

1. AI and Decision-Making: From Data to Policy Insights

2. Estonia's e-Residency Program: Empowering Global Citizens with Digital Governance
3. Singapore's Smart Nation Initiative: A Case Study in AI-Driven Urban Planning
4. The Promise of AI-Driven Education Policies: The PISA AI Assessment Pilot Project
5. The Role of AI in Environmental Policy: The United Nations AI for Earth Initiative

AI-Powered Enforcement and Governance

1. Predictive Policing: AI in Crime Prevention and Law Enforcement
2. AI and Healthcare Policy: Enhancing Regulatory Compliance and Patient Safety
3. AI in Financial Regulation: The Role of Supervised Learning in Detecting Financial Crimes
4. AI-Powered Compliance and Audit: A Shift in Regulatory Oversight Paradigm

The Global Imperative: Policy Cooperation Beyond 2050

1. AI in Diplomacy: Navigating the Global Policy Landscape
2. The Power of Multilateralism: The Paris Agreement and Global Climate Policies
3. International Collaboration in AI Governance: The OECD's AI Principles
4. The Ethics of AI-Powered Policymaking: Toward a Global Consensus

Looking towards the future, we have an incredible opportunity to create policies more aligned with citizens' complex neurological behavior patterns. The convergence of neuroscience, AI, and policymaking presents an unprecedented chance to promote greater societal well-being and inclusivity. However, the transformative power of AI-driven policymaking also requires a thoughtful and coordinated approach to ethical considerations on a global scale. As Thomas Friedman's "The World Is Flat" approach emphasizes, policymaking should not be viewed through a narrow lens of national interests but rather as a collective responsibility towards a sustainable future, given the world's interconnectedness.

The use of AI in governance has shown potential in addressing complex global challenges, as seen in Estonia's digital governance, Singapore's smart city initiative, and the United Nations AI for Earth project. However, international collaboration and adherence to ethical guidelines are crucial to unlocking AI's benefits in policymaking and enforcement. By understanding the neurological factors that shape citizens' preferences and using AI to make informed decisions, public institutions can redefine policymaking for the 21st century. Additionally, a global cooperative spirit will be essential in addressing transnational challenges and creating a resilient and prosperous future beyond 2050.

We must embark on a shared journey to harness the transformative power of AI and neuroscience for a more equitable and sustainable world. The future is full of possibilities, and we must act now to paint it with the collective wisdom and vision. Let's explore the four critical aspects of AI-driven policymaking.

The Promise of AI

Understanding Citizens' Neurological Behavior Patterns

The Neuro politics Nexus

Unraveling the Connection Between Neuroscience and Policymaking; Advancements in neuroscience have deepened our understanding of how individuals process information, make decisions, and form preferences. This newfound knowledge is increasingly shaping policymaking to align with the intricate workings of the human brain. One real-life example of leveraging neuro-AI policies is *Nudging Health Behavior.*

Governments worldwide are now using behavioral economics to shape people's health choices. In the United Kingdom, the "nudge unit" (officially known as the Behavioral Insights Team) employs insights from neuroscience to encourage healthier behaviors among citizens. By subtly changing how information is presented or framing options, they steer people toward making better decisions. For example, simplifying food labels and placing healthier food options at eye level in supermarkets has improved consumers' dietary choices.

Cognitive Biases and Decision-Making

How Our Brains Influence Policy Preferences; Cognitive biases are inherent in human decision-making, often leading to suboptimal choices. Policymakers leverage this knowledge to design policies that account for and mitigate biases. A notable real-life example is *Retirement Savings Policies*.

In the United States, Singapore, and a few other countries, automatic enrollment in retirement savings plans has been a successful policy intervention based on understanding cognitive biases. Many individuals tend to procrastinate or feel overwhelmed when saving for retirement. Automatic enrollment leverages the "status quo bias," encouraging employees to save for retirement by default. As a result, participation rates have significantly increased, improving citizens' financial security during their golden years.

Leveraging Behavioral Economics

Nudging Policies to Align with Human Behavior; behavioral economics for social good explores how psychological factors influence people's decisions. By incorporating behavioral insights into policy design, governments can create interventions that guide citizens toward better outcomes. One real-life example is the Opt-Out Organ Donation Policy. Countries like Austria and Spain have implemented this policy, where citizens are assumed to agree to organ donation after death unless they opt out. This uses the "default bias," as people often choose the default option. Consequently, these countries have experienced a rise in organ donation rates compared to opt-in systems, which has allowed more organs to be available and more lives to be saved.

Neurotechnology and Public Policy

The Ethical Implications of Brain-Computer Interfaces; The emergence of brain-computer interfaces (BCIs) poses novel challenges and opportunities for policymakers. BCIs enable direct communication between the brain and external devices, potentially revolutionizing healthcare and accessibility for people with disabilities. Policymakers must navigate the ethical implications of BCIs to ensure their responsible and equitable use. A real-life example in this domain may be a *Brain-Computer Interface for Paralyzed Individuals*. Researchers have developed BCIs that enable paralyzed individuals to control robotic limbs, wheelchairs, or computer interfaces using their brain signals. Governments are facing questions about informed consent, data privacy, and ensuring access to BCIs for all individuals, including those from marginalized communities. Policy frameworks need to be established to strike a balance between promoting innovation and safeguarding users' rights.

Understanding the neurological behavior patterns of citizens has become a powerful tool for policymakers who aim to create more effective and inclusive policies. By utilizing neuroscience and behavioral economics insights, governments can design interventions that align with humans' decision-making processes. Real-life examples such as nudging health behavior, retirement savings policies, and opt-out organ donation demonstrate the positive impact of such policies on society. As neurotechnology advances, policymakers must consider the ethical implications of brain-computer interfaces to ensure their responsible and equitable integration into public policies. By combining scientific knowledge with policy expertise, governments can create policies that cater to the intricacies of human behavior, thus promoting well-being and prosperity for their citizens.

AI Driven Policymaking in The Flat world

Decision-Making From Data to Policy Insights

AI has emerged as a game-changer in policymaking, empowering governments to process vast amounts of data and extract valuable insights. By leveraging machine learning algorithms, policymakers can make data-driven decisions that optimize public services and address societal challenges. A significant real-life example of AI-driven policymaking may be *Singapore's Smart Nation Initiative*. Singapore's Smart Nation Initiative is a pioneering effort to harness AI and data analytics to create a seamless, efficient, and sustainable urban environment. The country employs AI algorithms to analyze real-time data from sensors embedded in various urban infrastructures, such as traffic lights and public transport systems. These insights enable policymakers to optimize traffic flow, reduce congestion, and enhance public transportation services, making Singapore a model for smart and sustainable cities.

Empowering Global Citizens with Digital Governance

Estonia, a frontrunner in digital governance, has taken AI-driven policymaking to new heights through its e-Residency program. This groundbreaking initiative enables non-Estonian citizens to become "e-residents" and access Estonian public services online, including starting and managing businesses remotely. The program utilizes AI-powered identity verification to ensure the security and authenticity of e-residents. By leveraging AI, Estonia has fostered a global community of entrepreneurs and innovators contributing to its digital ecosystem.

AI-Driven Education Policies

The Organization for Economic Cooperation and Development (OECD) is pioneering the integration of AI into education policies through the PISA AI Assessment Pilot Project. This initiative aims to assess student's abilities to interact with AI systems and understand AI's impact on society. The data collected will inform policymakers about the skills and knowledge required to navigate an AI-driven world. By integrating AI into education assessments, the OECD is shaping policies that equip students with the skills necessary for the future workforce.

The Promise of AI in Environmental Policy

The United Nations AI for Earth Initiative harnesses the power of AI to address pressing environmental challenges. AI-driven technologies, such as satellite imagery analysis and predictive modeling, assist policymakers in monitoring ecological changes, combating deforestation, and managing natural resources more efficiently. USING AI INSIGHTS, the UN can shape evidence-based environmental policies and support sustainable development goals worldwide.

AI-driven policymaking has transformed global governance, revolutionizing how governments collect, analyze, and apply data to create impactful policies. Real-life examples, such as Singapore's Smart Nation Initiative, Estonia's e-Residency program, the OECD's PISA AI Assessment Pilot Project, and the UN's AI for Earth Initiative, showcase the immense potential of AI in shaping various aspects of society.

The World Is Flat approach emphasizes learning from successful global examples and collaborating to build a shared vision for the future. As AI reshapes policymaking, international cooperation becomes increasingly critical in addressing cross-border challenges and fostering responsible and ethical AI adoption. By leveraging AI's potential to optimize public services, enhance education, and tackle environmental issues, policymakers can navigate the complexities of a connected world, working collectively to shape a sustainable and prosperous future beyond 2050.

AI-Powered Enforcement and Governance

AI in Crime Prevention and Law Enforcement

AI has revolutionized law enforcement by enabling predictive policing, a proactive approach to crime prevention. By analyzing historical crime data and patterns, AI algorithms can predict potential hotspots and deploy law enforcement resources strategically. A notable real-life example may be *Los Angeles' Predictive Policing Program*. The

Los Angeles Police Department (LAPD) implemented the PredPol system, an AI-powered predictive policing program.

The system analyzes historical crime data, such as locations, times, and types of crimes, to generate hotspot predictions for law enforcement officers. By deploying officers to these predicted hotspots, the LAPD has reported significant reductions in certain types of crime, leading to improved public safety.

However, this program has been terminated but has shown art of possible. Los Angeles Police Department's decision to terminate its controversial predictive policing program. The program, which relied on data analytics and algorithms to predict crime hotspots and potential criminal behavior, had faced significant criticism and concerns over its effectiveness and potential biases.

Amid growing calls for surveillance reform and increased scrutiny of law enforcement practices, the LAPD's move to end the predictive policing program marks a significant step towards addressing the ethical and privacy implications of such technologies. Critics argued that the program perpetuated biases and infringed on individuals' civil liberties, leading to its discontinuation.

The decision highlights the need for responsible and transparent use of AI and data analytics in law enforcement. As society grapples with the potential benefits and risks of predictive policing, the LAPD's action serves as a crucial reminder of the importance of balancing security and privacy concerns. This development reflects the ongoing conversation surrounding the responsible implementation of technology in policing and underscores the significance of continued surveillance reform efforts to ensure fair and equitable policing practices.

Enhancing Regulatory Compliance and Patient Safety

The capabilities of AI go beyond just law enforcement and have also found use in healthcare policy. By analyzing vast amounts of patient data, AI-driven systems can identify potential risks and behaviors and improve quality of life while ensuring regulatory compliance. The FDA is an excellent example of how AI can be used in medical device safety surveillance. The U.S. Food and Drug Administration (FDA) has implemented AI-powered surveillance systems to oversee the safety and effectiveness of medical devices. By reviewing real-world data, such as electronic health records and reports of adverse events, the FDA can detect potential safety concerns and take appropriate regulatory action. AI is essential in promoting patient safety and simplifying the regulatory process.

The Role of Supervised Learning in Detecting Financial Crimes

The financial industry and governments must deal with complex fraud and money laundering challenges. However, AI technology, specifically supervised learning algorithms, has proven extremely useful in detecting financial crimes by analyzing vast quantities of financial data. An excellent example of this is the HSBC's AI-powered AML Systems. This multinational bank uses AI algorithms to ensure compliance with anti-money laundering (AML). These systems can scrutinize large amounts of transactional data and identify suspicious patterns, flagging potential money laundering or illicit financial activities. Financial institutions can meet regulatory requirements and protect their customers from financial crimes using the latest AI technology.

AI-Powered Compliance and Audit

AI has dramatically improved the efficiency and effectiveness of compliance and audit functions. With the help of AI-driven tools, auditors can now analyze data on a larger scale, identify anomalies, and ensure compliance with regulations. A prime example of this is Deloitte's AI-Driven Audit Analytics. As a global professional services firm, Deloitte has integrated AI-driven analytics into its audit process. Using AI algorithms, the firm efficiently analyzes financial data, identifies potential errors or discrepancies, and enhances audit accuracy. This AI-powered approach allows auditors to conduct more comprehensive assessments while minimizing the risk of oversight.

AI has transformed how governments and organizations fight against crime, ensure healthcare compliance, and monitor financial activities. Real-life instances such as the LAPD's Predictive Policing Program, the FDA's AI-powered medical device safety surveillance, HSBC's AML systems, and Deloitte's AI-driven audit analytics demonstrate the unparalleled efficiency and accuracy of AI solutions.

As AI technologies continue to advance, it's crucial for policymakers and regulatory bodies to carefully consider ethical concerns, data privacy, and potential biases to ensure responsible and equitable use. The ability of AI to quickly process large amounts of data can be incredibly useful for enforcement agencies and auditors, enabling them to make more informed decisions that can ultimately improve public safety and regulatory compliance.

The World Is Flat approach emphasizes the global nature of these challenges, urging collaboration among nations to establish best practices, share knowledge, and collectively address ethical and regulatory complexities. By embracing the transformative potential of AI, we can create a more secure, transparent, and efficient governance landscape, benefiting citizens and organizations worldwide beyond the year 2050.

The Promise of AI

The Global Imperative and Policy Cooperation now and Beyond

Navigating the Global Policy Landscape with AI

In the era of interconnectedness, the use of AI in international diplomacy presents both opportunities and challenges. On the positive side, AI-powered diplomatic systems offer invaluable insights by analyzing vast amounts of data, enabling policymakers to make informed decisions and foster international cooperation. A notable example is China's AI-Driven Diplomacy, where the country has invested significantly in AI technologies to understand public sentiments and assess responses to its foreign policies through the analysis of international news and social media. By leveraging AI-driven insights, Chinese diplomats can enhance their engagements with other nations, ultimately bolstering China's global standing.

However, the integration of AI in diplomacy also raises concerns that need careful consideration. One potential drawback is the risk of perpetuating biases in AI algorithms, which may impact the objectivity and fairness of diplomatic decisions. Additionally, the use of AI in diplomacy may raise questions about privacy and data security, as sensitive information from various sources is analyzed. As with any transformative technology, the adoption of AI in diplomacy requires thorough analysis and oversight to ensure ethical and responsible use, balancing the benefits with potential pitfalls. While AI can undoubtedly bolster international diplomacy, it necessitates critical examination and thoughtful regulation to harness its full potential while safeguarding the values and interests of nations and individuals alike. Policymakers must be cautious in their approach, carefully analyzing the implications and addressing concerns to strike a balance between progress and responsibility in the realm of AI-driven diplomacy.

The Power of Multilateralism

The pressing issue of climate change has led to a growing recognition of the importance of collective global action. In this context, AI has emerged as a powerful tool in the fight against climate change, aiding in the analysis of climate data and the development of effective climate policies. The Paris Agreement and AI Climate Modeling exemplify the potential of AI-driven multilateralism in addressing climate challenges.

As a supporter of the treaty, one can appreciate the significance of The Paris Agreement, which has garnered participation from nearly every nation worldwide. With a shared commitment to limit global warming and reduce greenhouse gas emissions, the agreement represents a landmark effort to unite nations in combating climate change. AI-driven climate modeling plays a pivotal role in this endeavor, enabling more accurate predictions of climate patterns and identifying regions most vulnerable to the impacts of climate change. This information equips countries with valuable insights to devise tailored policies that align with shared climate goals, fostering global cooperation and collaboration.

On the other hand, a naysayer may raise concerns about the effectiveness of The Paris Agreement and the reliance on AI climate modeling. Skeptics might question the actual implementation and enforcement of the treaty's commitments by participating nations. Additionally, they might express reservations about the accuracy and reliability of AI climate models, as predicting complex climate patterns remains a challenging task. Naysayers could also highlight potential ethical implications related to data collection and privacy in AI-driven climate modeling.

Despite these differing perspectives, one cannot deny the potential of AI-driven multilateralism in addressing climate change. Supporters argue that with continued advancements in AI technology and a strong commitment to international cooperation, AI can serve as a valuable ally in the fight against climate change. Critics, on the other hand, stress the need for careful evaluation, transparency, and accountability in the use of AI in climate modeling and policymaking.

In this critical juncture, the global community must strike a balance between optimism and cautiousness, leveraging

AI's capabilities while acknowledging its limitations. Ultimately, the success of AI-driven multilateralism in combating climate change will depend on the collective efforts and determination of nations worldwide to translate data-driven insights into concrete, impactful actions for a sustainable and resilient future.

International Collaboration in AI Governance

As AI technologies continue to transcend borders, international collaboration in AI governance becomes increasingly crucial to ensure ethical and responsible AI adoption. Recognizing this need, the Organization for Economic Cooperation and Development (OECD) has taken the lead in developing the OECD AI Principles. This comprehensive framework provides member countries with guidelines to shape their AI policies, promoting transparency, fairness, and accountability in AI development and deployment.

A real-life example of such international collaboration can be seen through Canada's active participation in the Global AI Summit organized by the Saudi Data and AI Authority. This summit serves as a platform for nations to come together, exchange ideas, and collaborate on AI governance and ethics. Through these collective efforts, countries can collectively address AI's societal and economic implications, ensuring that AI technologies are deployed in a manner that upholds human rights, respects diversity, and fosters innovation.

The Global AI Summit provided a valuable opportunity for Canada and other nations to engage in meaningful discussions, share best practices, and build consensus on key AI governance issues. By collaborating with like-minded countries, Canada demonstrates its commitment to fostering a shared responsibility for responsible AI development. Through these international partnerships, governments can pool resources, expertise, and experiences, accelerating progress towards a global framework that prioritizes ethical and responsible AI adoption.

In the rapidly evolving landscape of AI technologies, international collaboration is a powerful tool to navigate the complex challenges and seize the immense opportunities presented by AI. By actively participating in initiatives like the Global AI Summit, Canada showcases its dedication to shaping an AI future that prioritizes the well-being of its citizens and the global community. Together, through collective efforts and shared commitments, nations can ensure that AI remains a force for positive change and a catalyst for sustainable development on a global scale.

The Ethics of AI-Powered Policymaking

The rapid advancement of AI has led to ethical concerns, which require a global agreement on AI governance. International cooperation is crucial to addressing these ethical challenges. The World Economic Forum has established a Global AI Council that brings together experts, policymakers, and industry leaders worldwide. The council's goal is to create policy guidelines and recommendations that promote AI technologies' ethical and responsible use. This collaborative effort encourages a shared vision of AI ethics, which inspires responsible AI development across borders.

AI has become a catalyst for international diplomacy, climate cooperation, and ethical governance in an increasingly interconnected world. Real-life examples, such as China's AI-driven diplomacy, the Paris Agreement's AI climate modeling, the OECD's AI Principles, and the World Economic Forum's Global AI Council, demonstrate how international collaboration in AI governance is shaping a sustainable and prosperous future.

The World Is Flat approach highlights the interdependence of nations and the need for collective action in addressing global challenges. AI transcends borders, and its responsible and ethical development requires shared values and guidelines. Through multilateral initiatives and collaborative efforts, nations can harness AI's transformative potential while safeguarding human values, promoting fairness, and advancing shared interests. Policymakers, governments, and international organizations must come together to forge a global consensus on AI governance, ensuring that AI technologies serve as a force for the collective good, driving us toward a future that extends beyond 2050 with prosperity, equity, and harmony.

Pioneering a Collective Vision for a Thriving Future

As we welcome the fascinating journey of the Neurological Revolution and AI-based governance, we find ourselves at a historic crossroads. Here, significant changes meet the infinite potential of human creativity. The discoveries made in understanding how consumers and citizens behave neurologically, the possibilities of AI-based policymaking, the advances in AI-based enforcement and governance, and the call for global policy cooperation beyond 2050 have paved the way for a thriving and sustainable future. I invite you to explore the intricate link between neuroscience and policymaking in neuro-policy and neuro-politics.

Armed with cognitive biases and behavioral economics knowledge, we have grasped the essence of nudging policies that harmonize with human behavior. With the power to influence health behavior, secure financial futures, and drive sustainable choices, our grasp on neuroscience has become a fulcrum of effective governance.

We explored the fascinating world of AI-driven policymaking, where data and intelligence merge to blur the boundaries between reality and imagination. We were amazed by the impressive achievements of Singapore's Smart Nation, Estonia's digital sovereignty, and the United Nations AI-powered initiative to create an environmentally resilient world. These examples have showcased the transformative power of AI. The message of Thomas Friedman's "The World Is Flat" inspired us to join forces as global pioneers and create a connected landscape of cooperation that extends far beyond 2050.

We catapulted ourselves into the frontlines of AI-powered enforcement and governance, where algorithms stand sentinel in safeguarding society's safety and prosperity. Witnessing the strides of predictive policing, AI-empowered healthcare regulation, and financial oversight, we recognize that the ethical rudder guides the course of AI's journey. In our pursuit of progress, a harmonious blend of technological innovation and moral compass becomes the North Star, leading us toward a world governed with responsibility and foresight.

The grand symphony of international cooperation and diplomacy enveloped us, beckoning us to embrace the universality of AI governance. Enthralled by China's AI diplomacy, the multilateral resolve of the Paris Agreement, and the spirit of the OECD's AI Principles, we realized that the solutions to our shared challenges transcend geographical boundaries. As the World Economic Forum's Global AI Council leads the charge toward an ethical AI ethos, we find ourselves dancing in unison, weaving the fabric of a sustainable and equitable future.

In the embrace of this conclusion, let us be resolute in our quest to embrace the Neurological Revolution and AI-driven governance. With the world as our stage and cooperation as our anthem, we venture forth, equipped with wisdom and resolve, to shape a tomorrow where neuroscience and AI elevate humanity to unparalleled heights. Together, we pen a new chapter where the indomitable spirit of collective vision propels us toward a thriving future where all nations, as one global community, thrive beyond 2050. Let us forge ahead, for in unity and foresight; we script a symphony of triumph that resonates through the ages.

The Promise of AI

PART 3

AI Ethics and Bias

The Promise of AI

Ethics and Unbiased Approaches

Welcome to the enlightening journey of exploring the profound relationship between citizens' neurological behavior patterns and the ethical imperative in AI policymaking. In this thought-provoking exploration, we delve into the intricate interplay between AI technologies, citizens' behaviors, and the critical role of governments and institutions in crafting ethical AI frameworks.

The rapid advancements in AI have ushered in a new era of possibilities and challenges for societies worldwide. As AI increasingly influences various aspects of our lives, it becomes crucial to comprehend how it interacts with citizens' neurological behavior patterns. This journey begins by examining the promise and challenges of adopting AI in federal and state agencies, specifically focusing on public safety, healthcare, education, transportation, financial services, and environmental protection.

We embark on a path of enlightenment as we uncover the profound implications of AI-driven policies on citizens' lives. Real-world case studies, like ProPublica's investigation into the COMPAS algorithm and Amazon's gender-biased hiring tool, offer invaluable lessons, unveiling the ethical dilemmas and biases that can emerge from AI implementation.

Our exploration expands further to encompass the importance of an ethical AI framework that empowers governments and institutions. From data ethics and transparency to bias mitigation, the privacy of citizens' data, human oversight, and accountability, we navigate the essential guidelines that guide ethical AI governance. We showcase how the GDPR sets a remarkable standard for data protection, exemplifying how ethical frameworks can safeguard individuals' privacy and empower them with data control.

Venturing into the future, we shine a light on how AI technologies will continue to impact society, emphasizing the urgency of setting policies and frameworks now. By embracing inclusivity, continuous auditing, and evaluation, we witness how governments can harness AI's transformative potential while ensuring fairness, justice, and respect for human rights.

In this prescriptive exploration, we extend an invitation to governments, institutions, and policymakers to craft ethical AI frameworks that resonate with human values and aspirations. By leveraging AI responsibly and mindfully, we empower ourselves to build a future that transcends beyond 2050, where AI-driven policies respect the inherent dignity of every citizen and safeguard their rights.

Join me as I unravel the profound connection between citizens' neurological behavior patterns and the ethical imperative in AI policymaking. Together, let us shape an AI future that envisions a harmonious coexistence between humans and intelligent technologies, profoundly impacting society for generations to come. The time for ethical AI governance is now, and this exploration paves the way for a visionary roadmap towards an equitable and just AI-driven world.

In the age of unprecedented technological advancements, the convergence of neuroscience, AI, and policymaking has unlocked new frontiers in understanding citizens' neurological behavior patterns. While these developments offer immense promise for societal progress, they also present profound ethical challenges that demand urgent attention. For millennia, biases based on gender, income level, ethnicity, and more have plagued societies, resulting in discrimination and injustice. Now, as AI permeates all aspects of governance, the risk of perpetuating and exacerbating these biases through AI-driven policies becomes a stark reality.

Let's embark on an illuminating journey to explore the intricate interplay between citizens' neurological behavior patterns and the ethics that have governed societies for millennia. We delve into the complexities of AI in policymaking, examining how failure to set robust frameworks can lead to the exploitation of citizens and the erosion of their rights. Moreover, we illuminate the imperative of safeguarding average citizens from falling victim to AI-based discrimination. By addressing the problem statement, assessing the current state, and presenting guidelines for an ethical AI framework, we aim to empower governments and institutions to embrace the responsibility of protecting the rights and dignity of all citizens.

The Ethical Tightrope of AI Policy-Making

As AI-driven decision-making becomes increasingly prevalent, concerns over biased and unethical outcomes loom large. AI algorithms rely on vast amounts of data to make predictions and decisions, often perpetuating existing societal biases present in the data. Biases based on gender, income, ethnicity, and other factors have deep historical roots, leading to systemic discrimination in various domains, such as hiring practices, lending decisions, and criminal justice. When AI systems learn from biased data, they inadvertently perpetuate these discriminatory patterns, amplifying existing inequalities.

Moreover, the opacity and complexity of AI algorithms pose challenges in explaining how decisions are made, leading to the "black box" problem. Citizens affected by AI-driven policies may be unable to comprehend or challenge the decisions that impact their lives. This lack of transparency erodes trust in governance systems and undermines the democratic principles on which modern societies are built.

Current State of AI Bias and Ethical Lapses in Policymaking

Real-world instances of AI bias and ethical lapses underscore the urgency of addressing these challenges. Numerous studies have revealed racial and gender biases in AI systems used in hiring, mortgage lending, and predictive policing. For example, AI-based hiring tools have been found to favor male candidates over equally qualified female candidates, perpetuating gender disparities in the workforce. In criminal justice, AI-driven risk assessment algorithms have shown racial bias, leading to harsher sentencing for minority individuals.

Furthermore, privacy concerns emerge as AI systems collect and analyze massive amounts of personal data. Without robust privacy regulations and safeguards, citizens' personal information can be exploited, leading to intrusive surveillance and potential breaches of data privacy.

Guidelines for an Ethical AI Framework for Governments and Institutions

To ensure that AI policymaking upholds ethical principles and protects citizens' rights, governments and institutions must adopt comprehensive frameworks. The following guidelines are essential components of an ethical AI framework:

Data Ethics and Transparency: Prioritize transparency in AI algorithms and data sources to understand how decisions are reached. Ensure data used for AI training is diverse, representative, and free from biases. Provide clear explanations of how AI decisions are made and allow citizens to question and challenge outcomes.

Bias Mitigation: Implement robust bias detection and mitigation techniques to prevent discriminatory outcomes. Develop AI models that identify and correct bias in data and decision-making processes. Continuously monitor AI systems for potential bias and recalibrate as needed.

Privacy and Data Protection: Establish stringent privacy regulations to safeguard citizens' personal information. Adopt privacy-by-design principles to ensure that AI systems are built with privacy considerations from the outset. Implement strict data access controls and encryption measures to prevent unauthorized access.

Human Oversight and Accountability: Ensure that AI systems do not operate autonomously and require human oversight. Establish mechanisms for accountability when AI systems yield incorrect or biased outcomes. Create avenues for citizens to appeal decisions made by AI systems.

Inclusive AI Development: Promote diverse and inclusive teams in AI development to prevent biased perspectives. Involve stakeholders from diverse backgrounds in shaping AI policies to avoid unintended negative consequences.

Continuous Auditing and Evaluation: Regularly audit AI systems to identify potential biases or ethical violations. Establish a mechanism for ongoing evaluation and refinement of AI policies to align with evolving ethical standards.

Forging an Ethical Path in AI Policymaking

As we stand at the cusp of a new era, the intersection of citizens' neurological behavior patterns and AI-driven governance beckons us to embrace ethical responsibility wholeheartedly. The problem statement of biased AI algorithms and ethical lapses calls us to action, recognizing the imperative of safeguarding average citizens' rights. By acknowledging the current state of AI bias and ethical challenges, we unveil the urgency to set forth comprehensive ethical AI frameworks.

Guided by transparency, bias mitigation, privacy protection, human oversight, inclusivity, and continuous evaluation, governments and institutions can forge an ethical path in AI policymaking. Drawing wisdom from case studies and references, we glean insights from real-world experiences to inform ethical practices.

In this transformative journey, the sacred responsibility of governments and institutions lies in crafting AI policies that safeguard the dignity and rights of every citizen. Embracing these guidelines, we empower AI to be an ally, not a foe, in advancing societal progress and equity. The future of AI policymaking is in our hands, and the ethical choices we make today will reverberate through generations, shaping a world where AI serves humanity's highest ideals with unwavering integrity.

The Promise of AI

Data Ethics and Transparency

As the deployment of AI-driven decision-making systems becomes increasingly widespread, concerns about bias and ethical lapses in policymaking have risen to the forefront of public discourse. Despite the potential benefits of AI in streamlining processes, optimizing resource allocation, and improving service delivery, these advancements come with inherent risks that demand immediate attention.

Bias in AI Algorithms: Amplifying Historical Inequities

One of the most significant challenges in AI policymaking lies in the potential perpetuation of historical biases present in the data used to train AI algorithms. AI systems learn from vast amounts of historical data, which can inadvertently encode biases related to gender, race, ethnicity, religion, and other protected characteristics. When these biases are integrated into AI models, they can lead to discriminatory outcomes in various domains, such as hiring, lending, and criminal justice.

For example, AI-driven hiring tools have been found to favor male candidates over equally qualified female candidates, reflecting long-standing gender disparities in the workforce. Similarly, risk assessment algorithms used in the criminal justice system have demonstrated racial bias, leading to harsher sentencing for individuals from minority communities. These examples underscore the urgent need to address bias in AI algorithms and prevent their propagation in policymaking.

The "Black Box" Problem: Lack of Transparency and Accountability

AI algorithms often operate as complex "black boxes," meaning that the decision-making process is not easily interpretable or understandable to the average citizen. This lack of transparency raises concerns about accountability and undermines public trust in AI-driven policies. Citizens affected by AI-based decisions may find it challenging to comprehend the reasoning behind those decisions, making it difficult to challenge or appeal unfavorable outcomes.

The "black box" nature of AI also poses challenges for policymakers and regulators in identifying and rectifying biased or unethical decisions. Without visibility into the internal workings of AI algorithms, it becomes challenging to pinpoint the root causes of bias and ensure accountability for any resulting harm or discrimination.

Privacy and Data Protection Concerns

AI systems rely heavily on vast amounts of data to make accurate predictions and decisions. This data often includes sensitive personal information, raising significant privacy concerns. Improper handling or unauthorized access to personal data can result in privacy breaches and potential misuse of information.

Furthermore, when AI algorithms operate on personal data, there is a risk of "overfitting," where the system becomes too specific to the training data and fails to generalize to new situations accurately. This can lead to biased outcomes, as the AI model may not adequately consider diverse perspectives and experiences.

Human-AI Interaction: Ethical Considerations in Decision-Making

In certain contexts, AI-driven decision-making may intersect with human decision-making, raising questions about the division of responsibility and ethical decision-making. Policymakers must grapple with how to strike a balance between the autonomy of AI systems and human oversight and accountability.

For instance, in autonomous vehicles, AI algorithms make critical decisions that impact human lives. Policymakers must address ethical dilemmas, such as how AI systems should prioritize the safety of passengers versus pedestrians in potential collision scenarios. Striking the right balance between the efficiency and accuracy of AI and the ethical values of society remains a significant challenge.

Unintended Consequences: Unforeseen Biases and Outcomes

AI-driven policies can have unintended consequences, leading to unforeseen biases and outcomes. Even well-intentioned AI models can inadvertently introduce bias due to unanticipated interactions within complex data. Policymakers must consider the potential for these unintended consequences and implement mechanisms to detect and address them promptly.

Navigating the Ethical Frontier

The current state of AI bias and ethical lapses in policymaking reveals the complexity of deploying AI-driven decision-making systems responsibly. As AI increasingly permeates various domains of governance, it is imperative to recognize and mitigate the risks associated with biased algorithms and opaque decision-making processes.

Addressing these challenges requires a multi-faceted approach that includes:

1. Ensuring diversity and inclusivity in AI development teams to minimize unintentional biases.
2. Implementing transparency measures to shed light on the decision-making process of AI algorithms.
3. Regularly auditing AI systems to identify and rectify biased outcomes.
4. Establishing privacy regulations and data protection measures to safeguard citizens' personal information.

5. Encouraging collaboration between policymakers, researchers, and technologists to collectively address ethical considerations.

By confronting these challenges head-on, policymakers can pave the way for AI-driven policies that align with ethical values and uphold the principles of fairness, justice, and inclusivity. The journey toward ethical AI policymaking demands vigilance, empathy, and a commitment to preserving the rights and dignity of every citizen in the era of AI-driven governance.

The Promise of AI

Bias Mitigation

Bias mitigation is a critical pillar of an ethical AI framework to ensure that AI-driven policies do not perpetuate or exacerbate existing societal biases. Bias in AI algorithms can lead to discriminatory outcomes, reinforcing historical inequities and compromising the integrity of decision-making processes. Governments and institutions must proactively address bias, implement robust mechanisms for detection and correction, and foster a culture of fairness and inclusivity in AI development and policy implementation.

Recognizing the Presence of Bias: Identifying Potential Pitfalls

The first step in bias mitigation is acknowledging the potential for bias in AI systems. Policymakers and developers must be vigilant in recognizing the presence of biases that may be embedded in training data or implicit in the algorithms. Raising awareness about the impact of bias and its far-reaching consequences is crucial in steering AI development towards ethical practices.

Diverse and Representative Training Data: Balancing Inputs

Minimizing bias starts with the quality and diversity of training data used to develop AI algorithms. Governments and institutions should ensure that data sets are representative of the population the AI system is meant to serve. By including data from various demographic groups and underrepresented communities, policymakers can balance inputs and reduce the risk of reinforcing existing biases.

Bias Detection and Evaluation: Continuous Monitoring

Regularly monitoring AI systems for potential biases is essential. Governments should conduct bias detection and evaluation, assessing the performance of AI algorithms across different demographics and user groups. This continuous monitoring helps policymakers identify and rectify any unintentional biases that may emerge during the AI system's deployment.

Fairness-Aware AI: Incorporating Fairness Metrics

Fairness-aware AI, which incorporates fairness metrics during algorithm development, is an essential tool in bias mitigation. Policymakers and developers can use fairness metrics to evaluate and optimize AI algorithms for fairness and equitable outcomes. By specifying fairness constraints during the model development process, policymakers can ensure that the AI system's decisions do not disproportionately impact certain groups.

Human Review and Oversight: Balancing Automation with Human Judgment

While AI algorithms can streamline decision-making processes, human review and oversight remain crucial in mitigating bias. Policymakers should implement mechanisms for human review to ensure that AI decisions align with ethical guidelines and do not lead to discriminatory outcomes. This balance between automation and human judgment enhances the transparency and accountability of AI-driven policies.

Iterative Improvement: Learning from Mistakes

Bias mitigation is an iterative process. Policymakers and developers should embrace a learning mindset and acknowledge that mistakes may occur. When biases are identified, governments and institutions should take corrective action, update AI models, and iteratively improve the system to minimize the recurrence of bias.

A Fair and Inclusive AI Future

Bias mitigation stands as a critical safeguard in ensuring that AI-driven policies uphold principles of fairness, equity, and inclusivity. Governments and institutions hold the responsibility of creating AI frameworks that transcend historical biases and promote ethical decision-making. By recognizing bias's presence, diversifying training data, continuously monitoring AI systems, incorporating fairness metrics, exercising human review, and embracing an iterative improvement approach, policymakers can forge an AI future that benefits all members of society.

In the journey towards ethical AI governance, the commitment to bias mitigation aligns with the moral imperative of treating all citizens with dignity and respect. An AI framework rooted in fairness empowers governments to build trust with their constituents, foster social cohesion, and steer AI-driven policies towards a future that champions equality, justice, and human rights. As governments and institutions champion bias mitigation, they move closer to a vision of AI that serves as a force for good, promoting the well-being and prosperity of all individuals in the ever-evolving landscape of AI-driven governance.

Privacy and Data Protection
Privacy of Citizens' Data

The protection of citizens' privacy is a fundamental pillar of an ethical AI framework. As AI-driven policies rely on vast amounts of personal data to make decisions, ensuring the privacy and security of this data is paramount. Governments and institutions must establish robust privacy regulations and data protection measures to safeguard citizens' personal information, foster trust, and uphold the principles of autonomy and consent in the AI era.

Privacy by Design: Incorporating Privacy from Inception

Privacy by Design should be a core principle in AI development. Governments and institutions must integrate privacy considerations from the very inception of AI projects. This approach ensures that privacy measures are baked into the design of AI systems, rather than being added as an afterthought. By embedding privacy as a default setting, policymakers demonstrate a commitment to safeguarding citizens' data from the outset.

Anonymization and Encryption

To protect citizens' data, governments and institutions should employ anonymization and encryption techniques. Anonymization ensures that personally identifiable information (PII) is stripped from datasets used for AI training, reducing the risk of data breaches. Encryption adds an extra layer of security, ensuring that data remains protected during transmission and storage. These techniques minimize the likelihood of unauthorized access to citizens' sensitive information.

Data Minimization

Practicing data minimization is essential to limit the amount of personal data collected by AI systems. Governments should collect and retain only the data necessary for the specific AI task at hand, reducing the exposure of citizens' information to potential risks. Adopting data minimization principles also ensures compliance with privacy regulations and protects citizens from unnecessary data profiling.

Informed Consent and User Control

Giving citizens control over their data is paramount to ethical AI governance. Governments and institutions should obtain informed consent from individuals before collecting and using their data for AI purposes. Transparent and easily understandable consent mechanisms empower citizens to make informed decisions about how their data is utilized. Additionally, allowing users to exercise control over their data, including the ability to access, correct, or delete it, reinforces the notion of individual agency and privacy rights.

Secure Data Sharing and Data Sharing Agreements

In scenarios where data sharing is necessary for the development of robust AI algorithms, governments and institutions should prioritize secure data sharing practices. Establishing data sharing agreements that outline the terms, purpose, and scope of data exchange ensures that data is used responsibly and in accordance with ethical guidelines. These agreements also provide a framework for accountability and oversight when sharing sensitive data.

Third-Party Audits and Oversight

To maintain public trust, governments and institutions should implement third-party audits and oversight mechanisms to verify compliance with privacy regulations and ethical AI standards. Independent audits provide an external evaluation of AI systems' data handling practices, ensuring that privacy is upheld and potential risks are identified and addressed.

Nurturing Trust through Privacy Protection

Privacy of citizens' data serves as a cornerstone of ethical AI governance, underpinning citizens' trust, and confidence in AI-driven policies. Governments and institutions must prioritize privacy by design, employ anonymization and encryption, practice data minimization, empower citizens with informed consent and user control, establish secure data sharing practices, and implement third-party audits and oversight to ensure compliance.

In upholding the privacy of citizens' data, policymakers demonstrate a commitment to respecting individual autonomy, protecting sensitive information, and fostering a relationship of trust between governments and their constituents. An AI framework built on strong privacy principles empowers citizens to embrace the benefits of AI-driven governance while safeguarding their rights and preserving the integrity of their personal information. As governments and institutions champion privacy protection, they forge a path towards an AI future that not only delivers efficiency and innovation but also embodies ethical principles that promote the dignity and rights of every citizen.

Human oversight and accountability

Human oversight and accountability are essential components of an ethical AI framework, ensuring that AI systems are guided by human values and ethics. While AI algorithms can streamline decision-making processes, it is crucial to strike a balance between automation and human judgment. Governments and institutions must implement mechanisms for human review, establish accountability frameworks, and foster a culture of responsibility to address potential biases and ethical considerations in AI-driven policies.

Hybrid AI Systems

To avoid undue reliance on AI algorithms, governments and institutions should adopt hybrid AI systems that blend automated decision-making with human oversight. Hybrid systems allow AI to make data-driven recommendations while empowering humans to make the final decisions. By combining AI's analytical capabilities with human judgment, policymakers can ensure that AI decisions align with ethical guidelines and reflect the broader societal context.

Explainable AI (XAI)

The adoption of Explainable AI (XAI) methodologies is vital in promoting human oversight and accountability. XAI techniques, such as Local Interpretable Model-agnostic Explanations (LIME) and Shapley values, offer interpretable explanations for individual AI predictions. By providing clear explanations of how AI systems arrive at specific decisions, XAI enables policymakers and stakeholders to validate AI outcomes, challenge potential biases, and maintain control over AI-driven policies.

Human-in-the-Loop Systems

Human-in-the-loop systems involve human reviewers actively participating in the decision-making process alongside AI algorithms. This approach allows humans to verify the accuracy of AI predictions, detect biases, and intervene when necessary. Human-in-the-loop systems create a feedback loop that continuously refines AI models and enhances the overall quality of decision-making.

Ethical Review Boards

Establishing ethical review boards or committees is crucial in ensuring independent oversight of AI-driven policies. These boards, composed of experts from diverse disciplines, evaluate AI algorithms and policies for potential ethical dilemmas and bias. Ethical review boards contribute to impartial evaluations and safeguard against undue influence or conflicts of interest.

Accountability Mechanisms

Policymakers should implement accountability mechanisms that hold AI systems and developers responsible for their actions. Transparent reporting on the use and outcomes of AI algorithms helps maintain public trust and allows citizens to understand the impact of AI-driven policies on their lives. Accountability also involves documenting the decision-making process, including the data used, the reasoning behind decisions, and any actions taken to address identified biases.

Redress Mechanisms

Incorporating redress mechanisms is essential to rectify errors and bias that may occur in AI-driven policies. Governments and institutions should establish channels for citizens to contest AI decisions, seek explanations, and request corrections when necessary. Redress mechanisms empower citizens to challenge unfair or discriminatory outcomes and promote a sense of agency in AI-driven governance.

Striking the Balance for Ethical AI Governance

Human oversight and accountability form the ethical compass that guides AI-driven policies. Governments and institutions must embrace hybrid AI systems, Explainable AI, human-in-the-loop approaches, and ethical review boards to ensure that AI decisions are underpinned by human values and judgments. Transparency, accountability mechanisms, and redress mechanisms bolster public trust and provide channels for citizen participation and input in AI governance.

By striking the right balance between automation and human judgment, policymakers can cultivate a culture of responsibility and ownership in AI development and deployment. Human oversight and accountability strengthen the ethical fabric of AI-driven governance, empowering governments to harness the transformative potential of AI while preserving human dignity, fairness, and social cohesion. As governments and institutions champion human oversight and accountability, they pave the way for an AI future that reflects the best of human aspirations and ensures a harmonious coexistence between humans and intelligent technologies.

Inclusive AI Development

Inclusive AI development is a critical imperative for governments and institutions seeking to build ethical AI frameworks. To ensure AI-driven policies benefit all members of society, policymakers must prioritize diversity and inclusivity in the development of AI systems. By embracing multidisciplinary perspectives, involving underrepresented communities, and fostering collaboration, governments can create AI solutions that reflect the diverse needs, values, and aspirations of the entire populace.

Multidisciplinary Teams

Inclusive AI development begins with forming multidisciplinary teams that bring together experts from diverse fields such as technology, social sciences, ethics, and human rights. These teams can offer a broad range of perspectives, ensuring that AI algorithms are developed with a holistic understanding of societal contexts and potential impacts.

Gender and Diversity Representation

Ensuring gender and diversity representation within AI development teams is crucial in preventing biases in AI algorithms. Including women and individuals from underrepresented communities in the development process helps minimize the risk of bias perpetuation and promotes equitable AI solutions that cater to a wide range of users.

Community Engagement

Engaging communities and stakeholders affected by AI policies is essential to creating inclusive AI solutions. Governments and institutions should solicit citizen input through public consultations, town hall meetings, and feedback mechanisms. By incorporating citizen perspectives, policymakers can address societal concerns, identify potential biases, and align AI-driven policies with the needs of the people they serve.

Accessibility and Universal Design

AI applications should be designed with accessibility and universal design principles in mind. Policymakers must ensure that AI systems are usable by all individuals, regardless of their physical, cognitive, or sensory abilities. This commitment to inclusivity expands the reach of AI solutions and ensures that no segment of society is excluded from benefiting from advancements in technology.

Responsible Data Collection

When collecting data for AI development, governments and institutions must take extra care to protect vulnerable populations. Special considerations should be given to data collection from minors, elderly individuals, and marginalized communities. Data handling practices should prioritize the privacy and security of personal information, particularly for those who may be at higher risk of exploitation or discrimination.

Public Awareness and Education

Promoting public awareness and education about AI is essential for fostering inclusivity. Governments should invest in initiatives that enhance AI literacy among citizens, providing opportunities for understanding AI concepts and their implications. Educating the public empowers individuals to make informed decisions about AI technologies and participate meaningfully in discussions around AI policymaking.

Embracing Diversity for Inclusive AI Governance

Inclusive AI development is the bedrock upon which ethical AI governance is built. By assembling multidisciplinary teams, promoting gender and diversity representation, engaging communities, ensuring accessibility, protecting vulnerable populations, and promoting AI literacy, governments and institutions can foster inclusivity in AI-driven policies.

In embracing diversity, policymakers demonstrate a commitment to equity and justice in the AI era. Inclusive AI development broadens the perspectives that shape AI algorithms, leading to solutions that benefit everyone and leaving no one behind. As governments and institutions champion inclusive AI development, they create an AI future that embraces the collective genius of diverse minds, transcends barriers, and truly serves the greater good of humanity in the era of intelligent technologies.

Continuous Auditing and Evaluation

Continuous auditing and evaluation form a critical component of an ethical AI framework, ensuring ongoing scrutiny and improvement of AI-driven policies. As AI algorithms evolve and adapt to changing circumstances, it is essential for governments and institutions to implement mechanisms for continuous monitoring, auditing, and evaluation. By conducting regular assessments, policymakers can identify potential biases, detect unintended consequences, and respond proactively to emerging ethical challenges.

Real-time Monitoring

Real-time monitoring of AI systems allows governments and institutions to track the performance of AI algorithms continuously. This enables policymakers to detect potential biases or discrepancies as they occur and intervene promptly. By leveraging monitoring tools and analytics, policymakers can address biases before they lead to discriminatory outcomes and mitigate risks to public trust.

Algorithmic Impact Assessment

Conducting algorithmic impact assessments is crucial in evaluating the performance of AI algorithms across diverse user groups. Governments and institutions should regularly assess AI-driven policies to determine their impact on different demographics and identify any disparities in outcomes. This assessment helps ensure that AI systems function equitably and deliver fair results for all citizens.

Data Quality Assurance

Maintaining data quality is essential for the reliability and fairness of AI algorithms. Governments and institutions should establish data quality assurance measures to verify the accuracy, relevance, and completeness of data used for AI training. Ensuring data quality minimizes the risk of biased outcomes caused by inaccuracies or inadequacies in the training data.

Model Explainability and Interpretability

AI models should be designed with explainability and interpretability in mind. Policymakers should prioritize models that offer clear explanations of their decision-making processes. This transparency allows policymakers and stakeholders to understand how AI systems arrive at specific decisions, identify potential biases, and verify the fairness of AI-driven policies.

Ethical Stress Testing

Conducting ethical stress testing involves simulating various scenarios to evaluate AI system behavior under different conditions. Governments and institutions should test AI algorithms in simulated environments to understand how they respond to novel situations or edge cases. Ethical stress testing helps uncover potential biases or unintended consequences that may not be apparent under normal operating conditions.

Leveraging Stakeholder Insights

Implementing feedback mechanisms allows governments to gather insights and input from stakeholders, including citizens, civil society organizations, and domain experts. Feedback mechanisms provide opportunities for stakeholders to share their perspectives on AI policies, flag potential ethical concerns, and offer suggestions for improvement. By embracing stakeholder feedback, policymakers demonstrate responsiveness and adaptability to evolving ethical challenges.

A Journey of Continuous Improvement

Continuous auditing and evaluation are indispensable for ensuring the ongoing ethical integrity of AI-driven policies. By monitoring AI systems in real-time, conducting algorithmic impact assessments, assuring data quality, prioritizing model explainability, stress testing for ethical scenarios, and embracing stakeholder feedback, governments and institutions embark on a journey of continuous improvement.

In committing to continuous auditing and evaluation, policymakers exhibit their dedication to upholding ethical standards and maintaining public trust in AI-driven governance. As AI algorithms evolve, so too must the mechanisms for evaluating their performance and ethical implications. With a proactive and vigilant approach, governments and institutions navigate the complexities of AI ethics and create a future where AI-driven policies remain accountable, equitable, and in harmony with the evolving needs and values of society.

Case Studies

In the rapidly advancing world of artificial intelligence (AI), real-world experiences serve as invaluable case studies, offering crucial insights into AI's impact on society. As we explore the intricacies of AI's application in various domains, we encounter stories that shed light on both its potential and its pitfalls. ProPublica's investigation into the COMPAS risk assessment algorithm within the criminal justice system unraveled the disconcerting truth about racial bias, underscoring the urgent need for auditing AI systems to ensure fairness and accountability. Similarly, Amazon's gender-biased hiring tool exposed the imperative of scrutinizing AI for inclusivity and impartiality in the design of recruitment tools. Meanwhile, the European Union's General Data Protection Regulation (GDPR) stands as a beacon of data protection legislation, empowering citizens to safeguard their personal data while holding organizations accountable for misuse. As we traverse the AI landscape, we also encounter Google's Project Magenta, an inspiring endeavor promoting diversity in AI-generated content through creative tools like music composition. These real-world experiences offer profound lessons and underscore the significance of navigating the AI frontier with caution, responsibility, and a commitment to fostering inclusive, ethical, and innovative AI technologies.

ProPublica's Investigation into COMPAS Algorithm

The investigation conducted by ProPublica into the COMPAS (Correctional Offender Management Profiling for Alternative Sanctions) algorithm serves as a compelling case study highlighting the importance of addressing bias and ethical considerations in AI-driven policies, particularly within the criminal justice system.

Background:

COMPAS is an AI-driven risk assessment tool widely used in the United States to predict the likelihood of individuals re-offending or violating the terms of their parole. The algorithm analyzes various factors, including criminal history, age, and socio-economic background, to generate a risk score for each defendant. Courts often use this score to inform decisions regarding pre-trial detention, sentencing, and parole eligibility.

The Investigation:

In 2016, ProPublica conducted an extensive investigation into the COMPAS algorithm's predictive accuracy and potential racial bias. The investigation analyzed the outcomes of over 10,000 defendants in Broward County, Florida, over a two-year period. The key findings of the investigation were troubling:

Racial Bias:

ProPublica found that the COMPAS algorithm exhibited racial bias, disproportionately misclassifying Black defendants as high-risk and white defendants as low-risk. This raised concerns about the algorithm's fairness and whether it perpetuated existing racial disparities within the criminal justice system.

Unfulfilled Predictions:

The investigation revealed that the COMPAS algorithm's predictions were not entirely accurate. In a significant number of cases, defendants classified as high-risk did not re-offend, while some low-risk defendants did. These inaccuracies underscored the complexity of predicting human behavior and the potential risks of relying solely on algorithmic assessments for critical legal decisions.

Implications and Lessons Learned:

The ProPublica investigation into the COMPAS algorithm has profound implications for the use of AI in the criminal justice system and beyond. It highlights several important lessons:

Transparent Evaluation:

The investigation emphasized the need for transparent evaluation and auditing of AI algorithms used in critical domains such as criminal justice. Transparency allows for the identification of potential biases and inaccuracies, fostering public trust and confidence in AI-driven decision-making.

Explainability and Accountability:

The COMPAS case study underscored the importance of explainability and accountability in AI algorithms. For AI systems to be ethically sound, stakeholders must understand how they arrive at their predictions and be able to hold the algorithms accountable for their outcomes.

Incorporating Diverse Perspectives:

The investigation's findings shed light on the significance of including diverse perspectives during the development and evaluation of AI algorithms. Diverse teams can help identify and address potential biases, ensuring AI systems are fair and equitable.

Human Oversight:

The investigation highlighted the limitations of relying solely on AI algorithms without human oversight. Human judgment and expertise play a crucial role in balancing the strengths of AI with human values and ethical considerations.

The ProPublica investigation into the COMPAS algorithm serves as a stark reminder of the ethical challenges and risks associated with AI-driven policies. Governments and institutions must learn from such real-world experiences and proactively address biases, ensure transparency, embrace diverse perspectives, and exercise human oversight in AI development and deployment. By doing so, they can create ethical AI frameworks that foster fairness, justice, and societal well-being in an age where AI increasingly shapes decision-making in critical domains such as the criminal justice system.

Amazon's Gender-Biased Hiring Tool

Amazon's foray into AI-based hiring tools serves as a compelling case study, illustrating the potential pitfalls of relying on AI algorithms in critical processes like recruitment. The case sheds light on the importance of scrutinizing AI systems for bias and underscores the need for inclusivity and diversity in the design and development of such technologies.

Background:

In the early 2010s, Amazon sought to automate and streamline its hiring process by developing an AI-powered tool that would assist in identifying top candidates. The algorithm was designed to evaluate resumes and applications, ranking candidates based on their qualifications, experience, and past job performance. The goal was to expedite the recruitment process and enhance objectivity in candidate assessment.

The Unintended Bias:

As the tool underwent testing and implementation, Amazon soon realized that the AI algorithm was exhibiting gender bias. The system had been trained on historical data, which predominantly included resumes of male candidates, as the tech industry had been male-dominated for years. Consequently, the algorithm learned to associate certain male-centric keywords, colleges, and previous job experiences with higher qualifications.

The AI tool began systematically downgrading resumes that included terms often associated with female candidates or non-traditional backgrounds, resulting in a clear gender bias. The algorithm effectively perpetuated the gender disparity that already existed in the tech industry.

Identifying and Addressing the Bias:

Upon detecting the bias, Amazon took swift action and abandoned the AI hiring tool altogether. The company realized that the algorithm's biases were unacceptable and ran counter to their commitment to diversity and inclusivity. The case study serves as a stark reminder of the potential consequences of relying on AI systems without proper scrutiny and safeguards.

Lessons Learned:

The Amazon case offers several crucial lessons for governments and institutions:

Data Quality and Representation:

The case underscores the significance of using high-quality, representative data when training AI algorithms. Biased data can lead to biased outcomes, reinforcing existing disparities and potentially harming underrepresented groups.

Inclusivity in AI Design:

Inclusivity should be a core principle in the design and development of AI systems. Diverse teams with varied perspectives can help identify potential biases and ensure that AI algorithms treat all candidates fairly and equitably.

Transparency and Explainability:

The Amazon case highlights the importance of transparency and explainability in AI algorithms. It is vital to understand how AI systems arrive at their decisions, especially in high-stakes processes like recruitment.

Human Oversight and Intervention:

Human oversight is critical in assessing AI outcomes and ensuring that algorithms do not perpetuate biases. While AI can assist decision-making, human judgment should remain central in sensitive areas such as hiring.

Amazon's gender-biased hiring tool serves as a cautionary tale, emphasizing the need for robust evaluation and auditing of AI systems for potential biases. Governments and institutions must learn from real-world experiences like this case study and implement rigorous measures to ensure that AI-driven policies and tools adhere to ethical guidelines, promote inclusivity, and contribute to a fair and just society. By prioritizing diversity in AI design, fostering transparency, and incorporating human oversight, governments can harness the transformative power of AI while upholding the principles of equality and non-discrimination in the age of intelligent technologies.

European Union's General Data Protection Regulation (GDPR)

The General Data Protection Regulation (GDPR), enacted by the European Union (EU) in 2018, represents a landmark piece of data protection legislation that has significant implications for the rights and privacy of EU citizens. The GDPR stands as a prime example of robust data protection laws that empower individuals to have control over their personal data and impose stringent obligations on organizations to ensure responsible and lawful data handling.

Background:

The GDPR was introduced to address the growing concerns over data privacy, security, and the increasing prevalence of data breaches and misuse. The regulation was designed to harmonize data protection laws across the EU member states and replace the Data Protection Directive of 1995, which no longer adequately addressed the complexities of the digital era.

Key Provisions and Impacts:

Consent and Data Control: The GDPR places a strong emphasis on obtaining informed and explicit consent from individuals before processing their personal data. Organizations must clearly explain the purposes for data collection and provide users with easy-to-understand consent options. Citizens have the right to withdraw their consent at any time, putting them in control of their data.

Data Subject Rights: The GDPR grants data subjects a range of rights, including the right to access their data, the right to rectify inaccuracies, and the right to be forgotten (i.e., the right to have their data erased under certain circumstances). These rights empower individuals to manage their personal information and hold organizations accountable for data accuracy.

Data Breach Notification:

The GDPR requires organizations to promptly notify data protection authorities and affected individuals in the event of a data breach that poses a risk to individuals' rights and freedoms. This provision enhances transparency and helps individuals take necessary actions to protect their data.

Data Protection Officers (DPOs):

Organizations handling large-scale or sensitive data must appoint a Data Protection Officer. DPOs are responsible for ensuring compliance with the GDPR, providing expert advice, and acting as a point of contact for data protection matters.

Cross-Border Data Transfers:

The GDPR establishes a framework for cross-border data transfers, ensuring that personal data of EU citizens is protected when transferred outside the EU.

Impact and Global Influence:

The GDPR's influence extends beyond the EU's borders. It has served as a model for data protection legislation worldwide and inspired other countries to enact similar laws to safeguard citizens' privacy. The regulation has prompted organizations globally to reevaluate their data protection practices and align with GDPR principles, even when handling EU citizens' data outside the EU.

The European Union's General Data Protection Regulation (GDPR) sets a high standard for data protection laws globally. Its emphasis on consent, data control, data subject rights, data breach notification, and accountability has significantly empowered citizens and reshaped the way organizations handle personal data. The GDPR demonstrates that comprehensive data protection legislation can strike a balance between safeguarding individuals' privacy rights and promoting responsible data usage by organizations. As governments worldwide continue to grapple with the challenges of the digital age, the GDPR stands as a testament to the importance of prioritizing data privacy and fostering a culture of respect for individuals' personal information in the AI-driven era.

Google's Project Magenta

Google's Project Magenta is an AI research initiative that exemplifies the significance of diversity and inclusivity in AI development, particularly in creative domains such as music composition. The project showcases how incorporating diverse training data and perspectives can lead to more inclusive AI-generated content and foster creativity in the AI era.

Background:

Project Magenta was launched by Google's Brain team in 2016 with the mission of exploring the intersection of AI and creativity. The project's primary focus is to develop AI tools that can assist with creative tasks, including music composition, image generation, and other artistic endeavors.

Promoting Diversity in AI-Generated Music:

One of the noteworthy aspects of Project Magenta is its emphasis on diversity in the training data used for music composition. Recognizing the potential biases and limitations of training AI models on a narrow dataset, the researchers behind Project Magenta sought to create a more comprehensive and inclusive musical training dataset.

Music Genre Diversity:

Project Magenta's training dataset incorporates music from a wide range of genres and styles, spanning classical, jazz, pop, electronic, and many others. By including a diverse selection of musical genres, the AI models learn to produce music that reflects a broader spectrum of artistic expression.

Multicultural Influences:

In addition to genre diversity, Project Magenta aims to incorporate multicultural influences and musical traditions from various regions and cultures worldwide. By doing so, the project seeks to avoid cultural biases and ensure that AI-generated music is sensitive to and appreciative of diverse musical heritages.

Collaboration with Artists:

Project Magenta collaborates with artists, musicians, and composers from diverse backgrounds to provide valuable insights and feedback during the AI development process. This partnership ensures that AI-generated music is shaped by human creativity and cultural understanding.

Impact and Implications:

Project Magenta's commitment to diversity and inclusivity in AI-generated music has several noteworthy implications:

Promoting Inclusive Creativity:

By incorporating diverse training data, Project Magenta seeks to foster creativity that goes beyond traditional boundaries and celebrates a wide range of artistic expressions.

Avoiding Bias and Stereotypes:

The project's focus on diversity helps mitigate the risk of AI-generated music perpetuating biases or stereotypes, as it is exposed to a broader range of musical influences.

Cultural Sensitivity:

Project Magenta's collaboration with artists and musicians from diverse cultures ensures that AI-generated music is culturally sensitive and respectful.

Google's Project Magenta serves as an inspiring case study, demonstrating the power of diversity and inclusivity in AI development, especially in creative pursuits like music composition. By prioritizing genre diversity, multicultural influences, and collaborations with artists, Project Magenta contributes to a more inclusive and culturally aware AI-generated music landscape. As AI continues to shape creative industries, embracing diversity in AI development becomes essential for nurturing an AI future that celebrates the vast array of human expression and creativity in the globalized world.

Embracing the Ethical Imperative

As we conclude our journey of understanding citizens' neurological behavior patterns and the ethical imperative in AI policymaking, we stand at the threshold of a transformative era. The insights garnered from this exploration paint a vivid picture of the opportunities and challenges that lie ahead as AI technologies become increasingly intertwined with our lives.

Our quest began with an exploration of the promise and challenges of adopting AI in government agencies, shedding light on its potential impact on public safety, healthcare, education, transportation, financial services, and environmental protection. Real-world case studies illuminated the ethical dilemmas that can arise when AI is harnessed without careful consideration of biases and human-centric values.

The heart of our journey centered on the critical role of governments and institutions in crafting ethical AI frameworks. We discovered that data ethics, transparency, bias mitigation, privacy of citizens' data, human oversight, and accountability are the cornerstones of responsible AI governance. The GDPR's exemplary data protection legislation demonstrated that we can create frameworks that respect individuals' rights and foster trust between citizens and AI technologies.

With urgency and foresight, we embraced the imperative of setting AI policies and frameworks today. We recognized that continuous auditing and evaluation are vital to ensure that AI systems remain accountable and aligned with human values. Inclusivity emerged as a guiding principle to avoid perpetuating biases and to promote a future that celebrates diversity and cultural sensitivity.

Our prescriptive journey has laid a visionary roadmap for governments and institutions, encouraging them to act responsibly and mindfully in the AI-driven era. The key takeaways have paved the way for crafting an AI future that transcends beyond 2050—a future where AI technologies align harmoniously with human values, enhance public welfare, and protect individual liberties.

As we look ahead, the ethical imperative in AI policymaking becomes not just a necessity but a moral obligation. Embracing this imperative empowers us to harness AI's potential for the greater good, avoiding the pitfalls of unbridled technology. Together, we can create an equitable, just, and inclusive AI-driven world where citizens' neurological behavior patterns are respected, protected, and celebrated.

In the spirit of collaboration and collective wisdom, let us embark on this transformative journey—armed with ethics, innovation, and empathy—as we steer the course toward a future that empowers humanity, amplifies creativity, and redefines what it means to coexist with AI in the unfolding chapters of our shared history. The path is clear; the time for ethical AI governance is now.

The Promise of AI

Understanding Citizens Neurological Behavior Patterns

Unveiling the Ethical Imperative in AI Policy-Making

Objective: The framework aims to comprehensively explore the relationship between citizens' neurological behavior patterns and the ethical imperative in AI policymaking. It seeks to equip governments and institutions with the knowledge and guidelines necessary to develop ethical AI frameworks that prioritize citizen welfare, fairness, and respect for human rights.

Phase 1: Mapping AI Impact on Citizens' Behavior

1. Identify Key Sectors: Determine the government sectors where AI technologies are being adopted or planned, such as public safety, healthcare, education, transportation, financial services, and environmental protection.
2. Conduct Impact Assessment: Analyze the potential impact of AI adoption in each sector on citizens'

behavior, well-being, and rights. Identify potential biases, ethical challenges, and unintended consequences.

Phase 2: Ethics and Guidelines for AI Policy-Making

1. Data Ethics and Transparency: Establish data ethics guidelines for AI development, emphasizing transparency, informed consent, and clear explanations of AI decisions. Prioritize data quality and inclusivity in training datasets.
2. Bias Mitigation: Implement measures to mitigate biases in AI algorithms. Encourage diverse teams and collaborations with underrepresented communities to identify and rectify biases during AI development.
3. Privacy of Citizens' Data: Safeguard citizens' personal data through robust data protection laws and guidelines. Enable individuals to exercise control over their data and provide mechanisms for data rectification and erasure.
4. Human Oversight and Accountability: Emphasize the role of human judgment and expertise in AI decision-making. Establish accountability frameworks to address AI-generated outcomes and ensure adherence to ethical guidelines.
5. Inclusive AI Development: Promote the involvement of diverse perspectives in AI development, including gender, cultural, and socioeconomic diversity. Engage with stakeholders, experts, and artists to ensure cultural sensitivity and inclusivity.

Phase 3: Continuous Auditing and Evaluation

Real-time Monitoring: Implement real-time monitoring of AI systems to track performance and detect potential biases or anomalies. Enable prompt intervention to address ethical concerns.

Algorithmic Impact Assessment: Regularly assess the impact of AI-driven policies on different demographics to ensure fairness and identify disparities.

Ethical Stress Testing: Simulate ethical scenarios to evaluate AI system behavior under diverse conditions and edge cases. Uncover potential biases or unintended consequences.

Phase 4: Public Awareness and Education

AI Literacy Initiatives: Invest in public awareness and education programs to enhance AI literacy among citizens. Promote understanding of AI concepts, ethical implications, and citizen rights in the AI era.

Phase 5: Global Collaboration for Ethical AI Governance

International Cooperation: Foster collaboration and knowledge-sharing among governments and institutions to develop global standards for ethical AI governance.

Ethics Review Boards: Establish independent ethics review boards to evaluate AI projects and ensure adherence to ethical guidelines.

By adhering to this prescriptive framework, governments and institutions can embark on a transformative journey of understanding citizens' neurological behavior patterns and unveil the ethical imperative in AI policymaking. By prioritizing ethics, inclusivity, continuous auditing, and global collaboration, we pave the way for an AI future that respects human values, upholds citizen rights, and embraces diversity in the unfolding chapters of our shared history. The time to build an ethical AI-driven world is now, and this framework empowers us to navigate the path toward a future of equitable, just, and human-centric AI governance.

Part 4

The promise of AI for governments and citizens

The Promise of AI

The Promise of AI for Government

Artificial Intelligence (AI) is rapidly becoming a transformative force in various industries, and the government sector is no exception. In this chapter, we explore the immense promise AI holds for federal and state agencies, highlighting its potential to revolutionize governance and enhance public services. From streamlining administrative processes to augmenting decision-making capabilities, AI offers a plethora of opportunities for governments to operate more efficiently and effectively.

Importance of AI for Government:
1. Data-Driven Decision-Making: AI enables governments to harness vast amounts of data from diverse sources, facilitating evidence-based decision-making. This empowers policymakers to develop more targeted and impactful policies, addressing the specific needs of their citizens.

2. Enhanced Public Services: By leveraging AI technologies, governments can provide citizens with personalized and efficient services. Chatbots, for example, can offer 24/7 support, answering queries and guiding citizens through various procedures, ultimately leading to higher satisfaction rates.
3. Improved Efficiency and Cost Savings: AI automates repetitive and time-consuming tasks, reducing the bureaucratic burden on government agencies. As a result, resources can be allocated more effectively, leading to cost savings and optimized service delivery.
4. Predictive Analytics for Public Safety: AI can be employed to analyze crime patterns and predict potential threats, aiding law enforcement agencies in proactive crime prevention and ensuring public safety.
5. Personalized Education and Skill Development: AI-powered learning platforms can cater to the individual needs and learning pace of students, fostering a more inclusive and effective education system.

Key Takeaways:

1. AI enables data-driven governance, facilitating evidence-based policymaking and program evaluation.
2. Governments can enhance public services and citizen engagement through the implementation of AI-powered solutions, such as chatbots and personalized services.

3. Automation of administrative tasks with AI leads to increased efficiency and cost savings, enabling agencies to focus on critical issues.
4. AI's predictive capabilities empower law enforcement agencies to address public safety concerns more effectively.
5. AI-powered education platforms can revolutionize learning experiences, making education more accessible and tailored to individual needs.

Why Now is Critical to Start This Journey:

The current era presents a unique window of opportunity for government agencies to embark on the AI journey. Several factors contribute to the criticality of taking action now:

1. Technological Maturity: AI technologies have matured significantly, making them more accessible and easier to integrate into existing government systems. The time is ripe for agencies to leverage these advancements for better governance.
2. Rising Citizen Expectations: Citizens expect efficient and personalized services from their governments, akin to the seamless experiences the private sector provides. AI can help bridge this gap and elevate the quality of public services.
3. Global Competitiveness: Governments worldwide are embracing AI to enhance their competitiveness on the international stage. Federal and state agencies must adopt AI technologies to remain at the forefront of innovation and progress.
4. Data Abundance: The proliferation of digital data has created vast information repositories. AI's ability to analyze and extract valuable insights from

this data is invaluable for crafting informed policies and delivering targeted services.
5. Addressing Complex Challenges: Governments face multifaceted challenges, such as climate change, public health crises, and urbanization. AI's predictive and analytical capabilities can provide valuable assistance in addressing these complex issues.

I highlight the immense promise of AI for the government sector. It emphasizes the importance of leveraging AI to make data-driven decisions, enhance public services, improve efficiency, and ensure public safety. With AI technologies reaching maturity, rising citizen expectations, and the potential for global competitiveness, now is a critical moment for government agencies to embark on this transformative journey. By embracing AI, governments can revolutionize governance and create a more responsive and effective public sector, benefiting citizens and society.

The Challenges of Adopting AI in Government

As a seer in the Artificial Intelligence (AI) field, I understand AI's immense potential in transforming public services and governance. However, the journey towards adopting AI in the government sector has challenges. In this chapter, we will delve into the importance of AI for government agencies, identify key obstacles they may encounter, and present actionable strategies to overcome these challenges.

Importance of AI for Government:

Data-Driven Decision Making: AI empowers governments to harness the vast amounts of data they accumulate and turn it into valuable insights. Data-driven decision-making enables agencies to design evidence-based policies that effectively address citizens' needs.

Process Automation: AI-driven automation streamlines administrative processes, reducing human errors and freeing up valuable time for government employees to focus on higher-value tasks. This leads to increased efficiency and cost savings.

Enhanced Citizen Engagement: AI technologies, such as chatbots and virtual assistants, can provide citizens with instant support and answers to queries, ensuring a seamless and personalized user experience.

Predictive Analytics for Public Safety: AI-powered predictive analytics can identify patterns and trends in crime data, assisting law enforcement agencies in anticipating and preventing potential threats.

Improved Service Delivery: AI enables governments to optimize service delivery by tailoring services to individual needs, resulting in higher citizen satisfaction rates and increased trust in government institutions.

Key Takeaways:
1. Data-driven decision-making is the foundation of successful AI adoption in government, ensuring policies are based on accurate and relevant information.
2. Process automation with AI reduces administrative burden, leading to enhanced efficiency and resource optimization.
3. AI-powered citizen engagement tools improve public service delivery, fostering a positive perception of government services.
4. Predictive analytics empowers law enforcement to proactively address public safety concerns, leading to safer communities.

5. AI-driven personalized services cater to citizens' unique needs, enhancing overall satisfaction and trust in government agencies.

Challenges and Solutions:

Data Privacy and Security: Challenge: Government agencies deal with sensitive and personal data, making data privacy and security a primary concern. Solution: Implement robust data protection measures, adopt encryption technologies, and ensure compliance with data privacy regulations to safeguard citizen information.

Ethical AI Implementation: Challenge: Ensuring ethical AI deployment is essential to prevent bias and discrimination in decision-making processes. Solution: Establish clear AI ethics guidelines, conduct regular audits of AI algorithms, and prioritize fairness and transparency in AI models.

Workforce Readiness and Upskilling: Challenge: The adoption of AI may require the upskilling of government employees to effectively utilize AI technologies. Solution: Provide comprehensive training programs to equip the workforce with AI-related skills and encourage a culture of continuous learning.

Integration with Legacy Systems: Challenge: Integrating AI technologies with existing legacy systems can be complex and time-consuming. Solution: Develop a phased approach to AI integration, starting with pilot projects and gradually scaling up to ensure a smooth transition.

Examples:

Estonia's E-Residency Program: Estonia leveraged AI to simplify and streamline the process of applying for e-residency. This AI-powered system reduced the administrative burden and improved the overall experience for applicants.

Singapore's AI Governance Framework: Singapore developed an AI governance framework to ensure ethical and responsible AI use in the public sector. This initiative promotes transparency, accountability, and fairness in AI implementations.

Why Now is Critical to Start This Journey:

The current landscape presents a unique window of opportunity for government agencies to embark on the AI journey. Several factors contribute to the criticality of taking action now:

Rising Citizen Expectations: Citizens increasingly expect efficient, personalized, and technology-driven services from their governments. AI adoption can meet these expectations and enhance overall citizen satisfaction.

Global Competitiveness: Countries worldwide embrace AI in governance to gain a competitive edge. Governments that invest in AI now can position themselves as leaders in public service innovation.

Post-Pandemic Recovery: The COVID-19 pandemic highlighted the need for agile and tech-driven government responses. AI can be crucial in building resilient systems and responding effectively to future crises.

Technology Maturity: AI technologies have matured significantly, making them more accessible and easier to implement. Governments can leverage these advancements for immediate benefits.

This chapter we delves into the challenges of adopting AI in government and presents actionable solutions to overcome them. By recognizing the importance of data-driven decision-making, process automation, citizen engagement, predictive analytics, and personalized services, government agencies can position themselves for a transformative journey toward enhanced governance and public services. Now is a critical time for governments to embrace AI, ensuring they remain competitive, resilient, and responsive in a rapidly evolving world. As a consultant, I stand ready to support agencies on this transformative AI journey, driving positive change for citizens and society.

The Promise of AI

Empowering Tomorrow's Guardians

In the pursuit of safer communities and a more secure future, the incorporation of Artificial Intelligence (AI) in public safety is not just a lofty ambition but an urgent necessity. Previous chapters explore the paramount importance of AI in fortifying law enforcement, emergency response, and crime prevention. We uncover how AI-powered innovations can empower tomorrow's guardians to stay one step ahead of threats and crises, ultimately leading to a safer and more resilient society.

Importance of AI for Public Safety:

1. Proactive Crime Prevention: AI equips law enforcement agencies with predictive analytics, identifying patterns and anomalies in vast datasets to anticipate criminal activities and prevent them before they occur.

The Promise of AI

2. Real-time Emergency Response: AI-driven monitoring systems, combined with IoT devices, enable faster and more accurate identification of emergencies, facilitating rapid response and reducing response times during critical situations.
3. Enhanced Surveillance: AI-powered surveillance systems can intelligently monitor public spaces, automatically detecting suspicious behavior and providing real-time alerts to law enforcement personnel.
4. Intelligence-led Policing: AI assists in analyzing large-scale data to generate actionable insights, supporting intelligence-led policing strategies to target crime hotspots and allocate resources more effectively.
5. Bias Reduction and Fairness: Properly implemented, AI can minimize human biases in law enforcement decisions, promoting fairness and ensuring equal treatment of all individuals.

Key Takeaways:

1. AI's predictive analytics enhances crime prevention, enabling law enforcement to pre-empt potential threats and criminal activities.
2. Real-time emergency response with AI-driven monitoring systems saves crucial time during crises, potentially saving lives and minimizing damages.
3. AI-powered surveillance offers heightened security by identifying suspicious behavior and providing instant alerts to authorities.
4. Intelligence-led policing supported by AI analysis optimizes resource allocation and improves the effectiveness of law enforcement efforts.

5. Responsible AI implementation ensures fairness, reducing biases and promoting equitable treatment within public safety operations.
6. Real-World Examples: a. Chicago's Predictive Policing: The Chicago Police Department uses AI algorithms to predict crime hotspots, enabling officers to deploy resources strategically and deter criminal activities.
7. London's AI Video Surveillance: The Metropolitan Police Service in London utilizes AI-powered video surveillance to automatically identify and track suspicious behavior, enhancing public safety in crowded areas.

Why Now is Critical to Start This Journey:

The urgency to embrace AI in public safety stems from the escalating challenges faced by law enforcement agencies and emergency responders. Several factors accentuate the criticality of starting this transformative journey without delay:

1. Escalating Crime Rates: In an increasingly complex world, criminal activities are evolving rapidly. AI's predictive capabilities are crucial in staying ahead of emerging threats and preventing crime proactively.
2. Growing Population and Urbanization: Rapid urbanization increases the demands on public safety services. AI-driven systems can help manage the growing urban challenges, ensuring the safety of citizens in crowded environments.
3. Cybersecurity Threats: The rise of cybercrime poses unique challenges for public safety. AI can bolster cybersecurity measures, identifying potential cyber threats and protecting critical infrastructure.

4. Technological Advancements: AI technologies continue to evolve, and early adopters gain a competitive advantage in staying abreast of cutting-edge solutions for public safety.
5. Public Trust and Accountability: With increasing scrutiny of law enforcement practices, AI can help ensure accountability and transparency in policing, fostering public trust.

We underscore the imperative need for AI in public safety, highlighting its role in crime prevention, emergency response, and intelligence-led policing. By leveraging AI's predictive capabilities, real-time monitoring, and bias reduction, law enforcement agencies can usher in a safer and more secure future. The urgency to embrace AI lies in its potential to address contemporary challenges safeguarding communities against evolving threats and crises. As we embark on this transformative journey of AI for public safety, we must act now to equip our guardians with the tools they need to secure our future and uphold the values of a safer and more resilient society.

AI for Healthcare

In this chapter, we will explore the profound significance of Artificial Intelligence (AI) in revolutionizing the realm of healthcare. As we delve into the potential of AI-driven innovations, we envision a future where healthcare is not just about treating illnesses but about nurturing holistic wellness and personalized care. Join me on this transformative journey as we uncover how AI is set to pioneer the future of healthcare, bringing profound benefits to patients and healthcare providers alike.

Importance of AI for Healthcare:

1. Precision Diagnostics: AI's analytical prowess empowers healthcare professionals to interpret complex medical data with unparalleled precision. AI-driven diagnostics lead to earlier and more accurate detection of diseases, ensuring timely interventions and improved treatment outcomes.
2. Personalized Treatment Plans: With AI's ability to analyze vast patient datasets, healthcare providers can develop personalized treatment plans tailored to an individual's unique genetic makeup, lifestyle,

and medical history, fostering a new era of patient-centric care.
3. Predictive Healthcare Analytics: AI's predictive capabilities enable healthcare organizations to anticipate disease outbreaks, resource demands, and patient needs. This foresight enhances healthcare preparedness and resource allocation, especially during public health emergencies.
4. Drug Discovery and Development: AI expedites the drug discovery process by analyzing massive datasets and simulating complex molecular interactions. This facilitates the identification of potential drug candidates, accelerating the journey towards novel therapies.
5. Telemedicine Advancements: AI-driven telemedicine platforms optimize remote patient care, offering intelligent virtual consultations and remote monitoring to bridge the gap between patients and healthcare providers.

Key Takeaways:

1. AI-powered diagnostics ensure earlier disease detection, leading to more effective treatments and improved patient outcomes.
2. Personalized treatment plans revolutionize healthcare by tailoring therapies to individual patients, maximizing efficacy, and minimizing adverse effects.
3. Predictive healthcare analytics bolster preparedness for public health emergencies, enabling swift and effective responses.
4. AI expedites drug discovery, developing novel therapies and potential cures for a myriad of diseases.

5. Telemedicine advancements facilitate seamless remote patient care, bringing healthcare services closer to patients.
6. Real-World Examples: a. IBM Watson for Oncology: IBM Watson's AI platform assists oncologists in analyzing vast medical literature and patient records to recommend personalized cancer treatment options based on the patient's specific profile.
7. DeepMind's AI in Eye Disease Diagnosis: DeepMind's AI algorithms can detect early signs of eye diseases from optical coherence tomography (OCT) scans, helping ophthalmologists intervene before vision loss occurs.

Why Now is Critical to Start This Journey:

The urgency to embrace AI in healthcare springs from the convergence of several transformative factors:

1. Growing Healthcare Demands: An aging population and a rise in chronic diseases necessitate innovative solutions to meet the escalating demands on healthcare systems. AI-driven healthcare offers the scalability and efficiency needed to address these challenges.
2. Data Explosion: Healthcare generates an unprecedented amount of data daily. AI's ability to process and analyze this wealth of information presents an unparalleled opportunity to extract valuable insights for enhanced patient care.
3. Global Health Crises: Recent pandemics have exposed the vulnerabilities of healthcare systems worldwide. AI's predictive capabilities and telemedicine advancements equip healthcare

providers to respond effectively to crises and maintain continuity of care.
4. Technological Advancements: The rapid evolution of AI technologies makes now the opportune time to leverage AI's transformative potential in healthcare. Early adoption enables healthcare providers to stay at the forefront of medical advancements.
5. Patient-Centric Care: Patients expect personalized and holistic care tailored to their specific needs. AI-driven healthcare ensures that patients receive the attention and treatments that align with their unique medical profiles.

It propels us into the future of healthcare, where AI-driven innovations herald a new era of precision medicine, personalized care, and predictive healthcare analytics. By embracing AI's potential in diagnostics, treatment planning, drug discovery, telemedicine, and beyond, healthcare providers can revolutionize patient care and wellness. The time is now to embark on this transformative journey, redefining healthcare from reactive treatment to proactive prevention and holistic well-being. Together, let us usher in a future where AI and healthcare unite to nurture the health and happiness of humanity.

Empowering Lifelong Learning in a Transformative Era

Welcome to the future of education, where Artificial Intelligence (AI) emerges as a guiding light, transforming how we learn and acquire knowledge. In this visionary chapter, we embark on an enlightening journey into AI-driven education, where personalized learning experiences, adaptive curricula, and lifelong learning take center stage. Together, we explore the profound importance of AI in shaping the future of education, empowering individuals to thrive in a dynamic and ever-evolving world.

Importance of AI for Education:
1. Personalized Learning Pathways: AI's ability to analyze individual learning patterns and preferences allows for personalized learning pathways tailored to each student's unique strengths and weaknesses. This fosters a more engaging and effective learning experience.
2. Adaptive Learning Platforms: AI-powered adaptive learning platforms dynamically adjust the difficulty

and content of educational materials based on a student's progress, ensuring optimal knowledge retention and mastery.
3. Intelligent Tutoring Systems: AI-driven intelligent tutoring systems offer personalized and real-time feedback, enabling students to identify areas for improvement and receive timely guidance to enhance their learning journey.
4. Data-Driven Decision Making: Educators and administrators can leverage AI-generated insights to make data-driven decisions about curriculum design, resource allocation, and educational policy, optimizing learning outcomes for all.
5. Lifelong Learning Opportunities: AI-driven education extends beyond traditional classroom settings, providing accessible and continuous learning opportunities for individuals of all ages and backgrounds.

Key Takeaways:

1. Personalized learning pathways empower students to learn at their own pace, optimizing knowledge acquisition and fostering a love for learning.
2. Adaptive learning platforms offer tailored content, ensuring students are challenged appropriately and supported to achieve their full potential.
3. Intelligent tutoring systems provide personalized feedback, guiding students towards academic success and nurturing their confidence as learners.
4. Data-driven decision-making enables educators to make informed choices, refining educational approaches for greater effectiveness.
5. Lifelong learning opportunities facilitated by AI ensure that education remains a lifelong pursuit,

empowering individuals to adapt and excel in a rapidly changing world.

Examples:
1. Duolingo: AI-powered language learning platform Duolingo offers personalized lessons and adapts the difficulty level based on a student's proficiency, optimizing language acquisition.
2. Khan Academy: Khan Academy's adaptive learning platform adjusts the difficulty of math exercises based on a student's performance, providing targeted support and challenges as needed.
3. IBM's Watson Classroom: Watson Classroom utilizes AI to analyze student performance data, helping educators identify learning gaps and implement targeted interventions.
4. Coursera and Udemy: Online learning platforms like Coursera and Udemy utilize AI to recommend relevant courses based on a user's interests and learning history, promoting continuous education.

Why Now is Critical to Start This Journey:

The urgency to embrace AI in education stems from the evolving landscape of learning and the imperative to prepare students for a rapidly changing future. Several factors accentuate the criticality of starting this transformative journey now:

1. Global Digital Transformation: As technology permeates every aspect of society, students must be equipped with the digital skills and adaptability to thrive in a technology-driven world.

2. Lifelong Learning Imperative: In a future characterized by constant innovation and career shifts, AI-driven education enables individuals to remain competitive and relevant through continuous learning.
3. Diverse Learning Needs: Each student possesses unique learning preferences and strengths. AI's capacity for personalization ensures that educational experiences cater to the individual, promoting inclusivity and fostering a love for learning.
4. Real-time Feedback and Intervention: Timely feedback through AI-driven intelligent tutoring systems empowers students to overcome challenges promptly, reducing learning gaps and enhancing academic success.
5. Data-Driven Continuous Improvement: With AI-generated insights, educators can refine educational approaches, adapt curricula, and optimize resource allocation, ensuring the most effective and efficient learning outcomes.

In this chapter we embark on a transformative exploration of AI for education, where personalized learning, adaptive platforms, and lifelong learning opportunities converge. Embracing AI in education is vital to prepare students for a future marked by constant change and digital innovation. Through personalized learning pathways, intelligent tutoring, and data-driven decision-making, we create an educational ecosystem that empowers learners of all ages to reach their full potential and become lifelong knowledge seekers. Now is the opportune moment to embark on this journey, as AI opens the doors to a bright future of education that empowers individuals to thrive in a dynamic and ever-evolving world. Together, let us illuminate the path towards a future of transformative learning and empowerment through AI-driven education.

The Promise of AI

Pioneering Smart Mobility in a Connected World

Embark on a journey into the future of transportation, where innovation and technology are actively transforming how we move. This visionary chapter will delve into the groundbreaking ways AI is revolutionizing transportation. From self-driving vehicles to intelligent traffic management and sustainable mobility solutions, we will explore AI's significant role in shaping the future of transportation. Our mission is to pave the way toward intelligent mobility in our increasingly connected world.

Importance of AI for Transportation:

1. Autonomous Vehicles: AI's advanced perception and decision-making capabilities enable the development of autonomous vehicles, promising safer and more efficient transportation while reducing accidents caused by human error.

2. Traffic Management and Optimization: AI-powered traffic management systems analyze real-time traffic data to optimize signal timings, reduce congestion, and enhance overall traffic flow, leading to reduced travel times and fuel consumption.
3. Connected Mobility: AI facilitates seamless connectivity between vehicles and infrastructure, creating a cooperative network that enhances safety, traffic efficiency, and the overall travel experience.
4. Predictive Maintenance: AI-driven predictive maintenance ensures proactive identification of potential vehicle issues, leading to reduced downtime, optimized fleet management, and enhanced passenger safety.
5. Sustainable Transportation Solutions: AI's data analytics and optimization capabilities support the development of eco-friendly transportation solutions, such as route planning to minimize carbon emissions and promote greener mobility.

Key Takeaways:

1. Autonomous vehicles promise safer and more efficient transportation, revolutionizing the future of mobility.
2. AI-driven traffic management optimizes traffic flow and reduces congestion, leading to faster travel times and improved fuel efficiency.
3. Connected mobility fosters a cooperative network that enhances safety and travel experience through seamless vehicle-to-infrastructure communication.
4. Predictive maintenance ensures optimal vehicle performance, reducing downtime and enhancing fleet management.

5. Sustainable transportation solutions driven by AI promote eco-friendly mobility, reducing environmental impacts and fostering a greener future.

Examples:

1. Tesla's Autopilot: Tesla's Autopilot feature exemplifies AI's potential in enabling semi-autonomous driving capabilities. By leveraging AI algorithms and real-time data, Autopilot enhances safety and driver convenience.
2. Waymo's Self-Driving Cars: Waymo, a subsidiary of Alphabet Inc., is a leading company in the autonomous vehicle space. Its self-driving cars utilize AI technologies, such as computer vision and machine learning, to navigate complex urban environments.
3. Singapore's Smart Traffic Management: Singapore employs AI-based traffic management systems to optimize traffic flow, reducing congestion in its urban areas and enhancing overall transportation efficiency.
4. UPS's Predictive Maintenance: UPS, a global logistics company, utilizes AI-driven predictive maintenance to monitor its extensive fleet of vehicles, ensuring optimal performance and timely maintenance.

Why Now is Critical to Start This Journey:

The urgency to embrace AI in transportation arises from the pressing need to transform mobility for the future. Several factors accentuate the criticality of starting this transformative journey now:

1. Safety Imperative: As road accidents remain a leading cause of fatalities globally, AI-powered autonomous vehicles offer the promise of safer roads and enhanced transportation security.
2. Technological Maturity: AI technologies have matured significantly, making them ripe for widespread implementation in the transportation sector. Now is the opportune moment to capitalize on AI's potential.
3. Climate Change and Sustainability: As environmental concerns take center stage, the transportation sector must prioritize sustainability. AI's ability to optimize routes, promote electric vehicles, and reduce carbon emissions is critical in this context.
4. Urbanization and Population Growth: The world is rapidly urbanizing, leading to increased transportation demands. AI-driven solutions are necessary to optimize urban mobility and accommodate growing urban populations.
5. Economic Advantages: Embracing AI in transportation enhances a country's competitiveness on the global stage, attracting investment and driving economic growth through innovative smart mobility solutions.

We explore the future of transportation and how AI-powered innovations are leading the way to autonomous vehicles, intelligent traffic management, and sustainable mobility solutions. By adopting AI in transportation, we are ushering in an era of autonomous vehicles, intelligent traffic systems, and connected mobility. The time is right to start this transformative journey as AI technologies reach maturity and global demands for greener and smarter transportation solutions rise. Together, let's create a future where AI is the driving force behind transportation, revolutionizing mobility and creating a safer, greener, and more connected world.

The Promise of AI

AI for Financial Services

Welcome to the future of financial services, where Artificial Intelligence (AI) reigns as a transformative force, reshaping the landscape of banking, investment, and customer experience. In this visionary chapter, we embark on a thrilling journey into AI-driven financial services, where intelligent algorithms, automated processes, and personalized offerings redefine how we interact with money. Together, we'll explore the profound importance of AI in shaping the future of finance, paving the way for a more inclusive, efficient, and customer-centric financial ecosystem.

Importance of AI for Financial Services:
1. Enhanced Customer Experience: AI-powered chatbots and virtual assistants provide instant and personalized support to customers, ensuring round-the-clock assistance and seamless interactions. This elevates customer satisfaction and loyalty, revolutionizing the banking experience.
2. Personalized Financial Planning: AI-driven analytics analyze vast amounts of customer data to

offer tailored financial planning and investment advice. These personalized recommendations empower individuals to make informed decisions about their financial future.
3. Fraud Detection and Security: AI's advanced pattern recognition capabilities bolster fraud detection systems, minimizing financial risks and ensuring the security of sensitive customer information.
4. Efficient Trading and Investment: AI algorithms analyze market trends and real-time data to optimize trading and investment decisions, providing financial institutions with a competitive edge.
5. Inclusive Financial Services: AI enables the automation of processes, reducing operational costs and enabling financial institutions to reach underserved populations, fostering financial inclusion.

Key Takeaways:

1. AI-driven chatbots and virtual assistants elevate customer experience, providing personalized support and enhancing customer satisfaction.
2. Personalized financial planning empowers individuals to make informed decisions about their financial future, driving financial literacy and empowerment.
3. AI-powered fraud detection systems enhance financial security, minimizing risks and safeguarding customer assets.
4. AI's data analytics optimize trading and investment decisions, providing a competitive edge to financial institutions.

5. Automation enabled by AI fosters financial inclusion, reaching underserved populations and expanding access to financial services.

Examples:

1. Bank of America's Erica: Erica is Bank of America's AI-powered virtual financial assistant, helping customers manage their finances, provide personalized insights, and assist with various banking tasks.
2. Wealth-front and Betterment: Robo-advisors like Wealth-front and Betterment utilize AI algorithms to provide automated and personalized investment advice to individual investors, making wealth management more accessible and cost-effective.
3. PayPal's AI Fraud Detection: PayPal leverages AI algorithms to analyze transaction patterns and detect potential fraudulent activities, ensuring the security of users' financial information.
4. JPMorgan Chase's Contract Intelligence: JPMorgan Chase employs AI-powered contract intelligence to review and analyze legal documents, reducing the time and resources required for contract management.

Why Now is Critical to Start This Journey:

The urgency to embrace AI in financial services arises from the convergence of several transformative factors:

1. Evolving Customer Expectations: In the digital era, customers expect seamless and personalized experiences in financial services. AI-driven

solutions meet these demands, enhancing customer satisfaction and retention.
2. Technological Advancements: AI technologies have reached a level of maturity that enables widespread adoption in financial services. Now is the ideal time to capitalize on these advancements.
3. Competitive Landscape: Financial institutions must remain competitive in an ever-evolving market. AI-driven solutions provide a competitive edge, enabling institutions to offer innovative and efficient services.
4. Financial Inclusion Agenda: The global push for financial inclusion requires innovative approaches to reach underserved populations. AI-driven automation reduces costs, making it easier to serve diverse demographics.
5. Regulatory Compliance: As financial regulations continue to evolve, AI solutions can streamline compliance processes, ensuring institutions adhere to regulatory requirements efficiently.

We discussed the transformative journey into the world of AI-driven financial services, where enhanced customer experience, personalized financial planning, and fraud detection redefine the financial landscape. By embracing AI, financial institutions revolutionize the way they interact with customers, offering personalized support and empowering individuals to make informed financial decisions. Now is the opportune moment to embark on this journey, as AI technologies mature, customer expectations evolve, and financial inclusion takes center stage. Together, let us shape a future where AI becomes the bedrock of financial services, fostering a more inclusive, efficient, and customer-centric financial ecosystem. As the financial world evolves, AI will continue to be the driving force, transforming banking, investment, and the way we engage with our finances. With AI as our ally, we embark on a journey towards a more financially empowered and connected future.

The Promise of AI

Building a Sustainable Tomorrow

Step into a world where Artificial Intelligence (AI) becomes a guardian of our planet, spearheading efforts to protect and preserve the environment. In this visionary chapter, we embark on an enlightening journey into AI-driven environmental protection, where advanced algorithms, predictive analytics, and data-driven insights shape a sustainable future. Together, we explore the profound importance of AI in safeguarding our planet, paving the way for eco-friendly practices, and addressing pressing environmental challenges.

Importance of AI for Environmental Protection:

1. Environmental Monitoring: AI-powered sensors and drones enable real-time monitoring of ecosystems, tracking changes in air and water quality, deforestation, and wildlife patterns. This data-driven approach enhances our understanding of environmental changes and supports timely interventions.
2. Climate Change Mitigation: AI algorithms analyze vast climate data sets to model and predict climate

patterns. These predictive analytics assist policymakers in formulating effective strategies for climate change mitigation and adaptation.
3. Wildlife Conservation: AI facilitates the monitoring and tracking of endangered species, helping conservationists identify and address threats to wildlife habitats. AI-powered image recognition technologies aid in wildlife identification and population estimation.
4. Sustainable Agriculture: AI-driven precision farming optimizes resource use, including water and fertilizers, increasing crop yields and reducing environmental impacts. This promotes sustainable agricultural practices and supports food security.
5. Waste Management: AI technologies enhance waste sorting and recycling processes, improving waste management efficiency and reducing landfill usage. AI-driven approaches also help identify opportunities for upcycling and waste-to-energy solutions.

Key Takeaways:

1. Environmental monitoring with AI enables real-time tracking of changes in ecosystems, supporting timely interventions for conservation.
2. AI's predictive analytics aids in climate change mitigation, assisting policymakers in formulating effective strategies.
3. AI plays a crucial role in wildlife conservation, aiding in species monitoring and habitat preservation.
4. Precision farming powered by AI promotes sustainable agriculture, optimizing resource use and increasing crop yields.

5. AI-driven waste management enhances recycling efficiency and identifies opportunities for sustainable waste solutions.

Examples:
1. Conservation Drones: Conservationists use AI-powered drones to monitor wildlife and detect illegal activities in protected areas, aiding in the protection of endangered species and habitats.
2. Climate Prediction: AI-powered climate models, such as those developed by climate scientists and organizations like NASA, help predict extreme weather events and support climate change adaptation measures.
3. Plantix: The Plantix app employs AI image recognition to diagnose plant diseases, empowering farmers with timely advice on pest control and sustainable farming practices.
4. IBM's Green Horizons: IBM's Green Horizons initiative utilizes AI to model and predict air pollution patterns, helping cities and industries make informed decisions to reduce their environmental footprint.

Why Now is Critical to Start This Journey:
The urgency to embrace AI for environmental protection arises from the pressing need to address the challenges posed by climate change and habitat destruction. Several factors accentuate the criticality of starting this transformative journey now:

1. Accelerating Climate Crisis: Climate change is already having far-reaching impacts on our planet,

necessitating immediate action and innovative solutions to mitigate its effects.
2. Advanced Data Analytics: AI technologies have advanced to a point where they can effectively analyze vast environmental data sets, offering valuable insights for decision-making.
3. Biodiversity Loss: Rapid biodiversity loss calls for innovative approaches to wildlife conservation. AI-powered monitoring can aid in preserving endangered species and their habitats.
4. Sustainable Development Goals: The global commitment to the United Nations Sustainable Development Goals underscores the need for urgent and collective efforts to protect the environment.
5. Collaborative Global Efforts: Governments, organizations, and industries worldwide are increasingly recognizing the importance of AI in environmental protection, creating momentum for transformative change.

In this chapter we discussed the transformative journey into the world of AI-driven environmental protection, where real-time monitoring, predictive analytics, and sustainable practices pave the way for a greener and more sustainable tomorrow. By embracing AI, we empower ourselves to be better stewards of our planet, addressing pressing environmental challenges and safeguarding the well-being of future generations. Now is the opportune moment to embark on this journey as AI technologies mature and the need for innovative environmental solutions intensifies. Together, let us shape a future where AI becomes the driving force behind environmental protection, fostering a more sustainable and ecologically conscious world. With AI as our ally, we embark on a journey towards a greener and more harmonious coexistence with nature. As the protectors of our planet, AI empowers us to build a sustainable tomorrow where we live in harmony with the Earth and its inhabitants.

The Promise of AI

Crafting a Tailored AI Vision for Your Department

As we near the end of our book, it's time to unleash the true potential of AI in your world with a prescriptive approach. In this chapter, we'll empower you to take charge, building your AI framework tailored to your business needs. With the knowledge and inspiration gained from our AI Adventures, you have the tools to embark on an exciting quest where innovation and AI-driven transformation await!

AI has demonstrated its versatility and potential across various industries, from healthcare and finance to education and transportation. Now, it's time to unleash this powerful force in your organization, uncovering unique opportunities and crafting solutions that push the boundaries of what's possible.

Let's embark on this exciting endeavor by taking inspiration from our AI companions who've made remarkable strides in their respective fields:

1. AI for Customer Experience: Channel the prowess of Bank of America's Erica or IBM's Watson, and create an AI-powered virtual assistant that wows your customers. Personalized interactions, 24/7 support, and intuitive recommendations will elevate your customer experience to new heights.
2. AI for Finance and Investments: Draw inspiration from robo-advisors like Wealth-front and Betterment, and build an AI-driven investment platform that offers tailor-made financial advice. Your customers will marvel at the precision of AI-powered portfolio recommendations and personalized wealth management.
3. AI for Healthcare: Follow in the footsteps of IBM Watson for Oncology and Plantix and develop an AI system that aids medical professionals in diagnosing complex diseases or assists farmers in identifying plant ailments. The power of AI in healthcare extends beyond diagnosis; it can optimize treatment plans and even aid in drug discovery.
4. AI for Education: Embrace the spirit of Duolingo and Khan Academy and create an AI-driven educational platform that adapts to individual learning needs. Personalized lesson plans, interactive quizzes, and progress tracking will make learning an exciting and rewarding journey for your students.
5. AI for Transportation: Join the league of Tesla's Autopilot and Singapore's Smart Traffic Management and pioneer smart mobility solutions. Develop autonomous vehicle technology or implement AI-powered traffic management

systems to revolutionize transportation in your region.

But remember, the key to a successful AI framework lies in understanding your organization's unique needs and challenges. It's not about replicating existing AI solutions but crafting an adventure that aligns with your business goals.

Here are a few steps to guide you on your AI quest:
1. Identify Business Challenges: Pinpoint the areas where AI can make the most significant impact in your organization. Whether optimizing processes, enhancing customer experiences, or predicting market trends, AI has a solution.
2. Data, the Treasure Trove: Gather and curate high-quality data to fuel your AI adventure. AI algorithms thrive on data, and the better the data, the more accurate and insightful the results.
3. Collaboration and Learning: Cultivate a culture of AI learning and collaboration within your organization. Encourage employees to explore AI's potential and provide opportunities for upskilling and training.
4. Partnerships and Expertise: Forge alliances with AI experts and solution providers who can complement your AI journey. Leverage their expertise to accelerate your AI initiatives.
5. Ethical AI: As you build your AI framework, ensure ethical considerations are at the heart of your AI adventures. Prioritize transparency, fairness, and responsibility in your AI implementations.

Dear readers, the possibilities are limitless, and the AI landscape is ever-evolving. Embrace the spirit of curiosity, innovation, and a daring sense of adventure as you embark on this AI quest. The future awaits, and with your very own AI framework, you'll be at the forefront of the next wave of transformation.

Remember, AI isn't just a tool; it's your steadfast ally, empowering you to create a future that transcends boundaries and unlocks the full potential of your business.

AI Framework for Policy Makers

Developing an AI framework is indeed of paramount importance once the vision for ethical AI governance has been established. A well-defined AI framework provides the necessary structure and guidelines to navigate the complexities of AI development and deployment. It outlines the principles, policies, and procedures that govern AI systems' design, implementation, and usage, ensuring alignment with ethical objectives. A robust AI framework emphasizes transparency, accountability, fairness, and inclusivity, setting the stage for responsible AI innovation. It incorporates mechanisms for data ethics, bias mitigation, privacy protection, and human oversight, fostering public trust and confidence in AI technologies. Furthermore, an AI framework should be adaptable and continuously evaluated, allowing for agility in response to emerging ethical challenges and advancements in AI research. By embracing an AI framework rooted in ethical principles and forward-thinking policies, governments can embrace the potential of AI while safeguarding citizens' rights and welfare, laying the groundwork for an inclusive and human-centric AI future.

Ensuring Ethical AI Governance

Objective: This framework is designed to guide government policy makers in developing ethical AI governance frameworks that prioritize human rights, fairness, accountability, and societal well-being. It draws inspiration from the OECD's "Principles on Artificial Intelligence" and aims to provide a practical and comprehensive approach to navigate the complexities of AI driven policymaking.

Phase 1: Setting Ethical Foundations

1. Engaging Stakeholders: Involve diverse stakeholders, including citizens, industry experts, academics, and civil society organizations, to ensure a comprehensive understanding of ethical concerns and perspectives.
2. Adopting Ethical Principles: Embrace the OECD's AI Principles as guiding tenets, emphasizing human-centered AI, fairness, transparency, accountability, and inclusivity.
3. Defining Ethical Goals: Set clear ethical objectives that prioritize the protection of human rights, privacy, and societal values in AI development and deployment.

Phase 2: Ensuring Transparency and Explainability

1. Transparency in AI Systems: Establish regulations that mandate the disclosure of the use of AI systems to users, ensuring transparency about AI's presence and impact on their interactions.
2. Explainable AI: Encourage the adoption of AI systems that provide transparent explanations of

their decisions, enabling users to understand how AI arrives at specific outcomes.

Phase 3: Fostering Fairness and Accountability

1. Bias Mitigation: Develop guidelines to address bias in AI algorithms, ensuring that AI-driven policies do not unfairly discriminate against certain individuals or groups.
2. Human Oversight and Accountability: Institute mechanisms for human oversight and intervention in critical decision-making processes, ensuring that AI is used as a supportive tool rather than an autonomous decision-maker.

Phase 4: Safeguarding Privacy and Data Protection

1. Data Governance and Consent: Implement robust data governance policies that protect citizens' data and require explicit consent for data collection, storage, and processing.
2. Data Security and Anonymization: Ensure that AI systems handle data securely and adopt anonymization techniques to protect individuals' identities.

Phase 5: Promoting Inclusivity and Accessibility

1. Inclusive AI Design: Advocate for AI systems that cater to the diverse needs of users, including individuals with disabilities and those from different cultural backgrounds.
2. Addressing Digital Divide: Bridge the digital divide by promoting access to AI technologies and ensuring that AI benefits are equitable across all segments of society.

Phase 6: Continuous Evaluation and Adaptation
1. Ethics Review Boards: Establish independent ethics review boards to evaluate AI projects and assess their compliance with ethical guidelines.
2. Continuous Auditing and Evaluation: Implement ongoing audits of AI systems to identify and address ethical concerns, ensuring that AI-driven policies remain aligned with societal values.

By adhering to this proposed comprehensive framework, policy makers can lay a strong foundation for ethical AI governance. Prioritizing human rights, transparency, fairness, and inclusivity in AI policymaking will foster public trust, promote innovation, and ensure AI's positive contribution to society. The journey towards ethical AI governance requires continuous efforts, collaboration, and adaptability to navigate the evolving challenges of the AI-driven era. With this framework as a guiding compass, governments can proactively shape an AI future that upholds human values, safeguards citizens' rights, and fosters a society where technology serves as a tool for the greater good.

Part 5

Envisioning Tomorrow's Possibilities

Art of possible

As AI technology continues to advance, its transformative potential in shaping the future of government and public services becomes increasingly apparent. One specific use case lies in the realm of citizen engagement and participation. AI-powered chatbots and virtual assistants can serve as accessible and responsive interfaces, enabling citizens to interact with government agencies effortlessly. These AI-driven platforms can provide personalized information, answer queries, and guide citizens through various processes, enhancing public service accessibility and efficiency. Moreover, AI's data analytics capabilities offer governments valuable insights into citizens' needs and preferences, enabling evidence-based policymaking and targeted service delivery. Additionally, AI can revolutionize decision-making processes by analyzing vast datasets and predicting trends, helping governments identify potential challenges and opportunities for sustainable development. From optimizing transportation routes to predicting public health outbreaks, AI-driven predictive analytics can empower governments to proactively address societal issues and foster long-term sustainable growth. Embracing AI technologies in public services and governance empowers governments to forge closer connections with citizens, deliver more tailored and effective services, and build a future where AI supports equitable and sustainable development.

1. AI-Driven Healthcare Services: AI-powered diagnosis and treatment recommendation systems can assist healthcare professionals in providing more accurate and personalized medical care. (Reference: "AI in Healthcare: Through the Lens of AI Experts," PwC, 2021)
2. Predictive Policing: Using historical crime data and AI algorithms, law enforcement can predict crime hotspots, enabling proactive measures to prevent criminal activities. (Reference: "Predictive Policing in Los Angeles: An Evaluation of the LASER Program," RAND Corporation, 2019)
3. AI-Assisted Disaster Response: AI can analyze real-time data during disasters to assess the impact and aid in effective disaster response and resource allocation. (Reference: "AI for Disaster Response and Recovery," United Nations Global Pulse, 2020)
4. AI-Powered Social Services: AI chatbots can offer immediate support to citizens seeking social services, providing timely information and assistance. (Reference: "AI and Government Social Services," Center for Data Innovation, 2022)
5. Smart Traffic Management: AI-enabled traffic management systems can optimize traffic flow, reduce congestion, and enhance road safety. (Reference: "AI in Urban Mobility," McKinsey & Company, 2021)

6. AI-Driven Tax Compliance: AI algorithms can assist tax authorities in detecting tax evasion and ensuring compliance with tax regulations. (Reference: "AI in Tax Compliance," World Bank, 2020)
7. AI-Enhanced Education Systems: AI-powered personalized learning platforms can adapt to individual students' needs, improving learning outcomes. (Reference: "AI in Education: Teaching and Learning in the Age of Intelligent Machines," OECD, 2019)
8. AI for Agricultural Productivity: AI-driven precision agriculture can optimize resource use, enhance crop yields, and promote sustainable farming practices. (Reference: "AI in Agriculture," Food and Agriculture Organization, 2022)
9. Virtual Government Assistants: AI chatbots can assist citizens with routine inquiries, providing 24/7 access to government services and information. (Reference: "Conversational AI in Government," National Association of State Chief Information Officers, 2021)
10. AI-Powered Public Safety Drones: Drones equipped with AI-powered image recognition can aid in search and rescue operations during emergencies. (Reference: "AI for Public Safety: Leveraging Drones," The Police Foundation, 2018)
11. AI-Enabled Energy Management: AI can optimize energy distribution and consumption, reducing waste and enhancing energy efficiency. (Reference: "AI in Energy Management," International Energy Agency, 2021)
12. AI-Enhanced Cybersecurity: AI-driven cybersecurity systems can detect and prevent cyber

threats more effectively, safeguarding critical government data. (Reference: "AI in Cybersecurity," World Economic Forum, 2020)
13. AI for Environmental Monitoring: AI-powered sensors and drones can monitor environmental changes and support conservation efforts. (Reference: "AI for Environmental Monitoring," United Nations Environment Programme, 2022)
14. AI-Driven Financial Regulation: AI algorithms can analyze financial data to detect anomalies and potential risks, aiding regulatory bodies in ensuring financial stability. (Reference: "AI in Financial Regulation," Bank for International Settlements, 2021)
15. AI-Enhanced Citizen Engagement: AI can analyze citizen feedback and sentiments to inform policy decisions and improve government services. (Reference: "AI for Citizen Engagement," GovLab, 2019)
16. AI for Language Translation: AI-powered language translation tools can facilitate communication and information exchange between diverse language speakers. (Reference: "AI Language Translation: A State-of-the-Art Review," Association for Computational Linguistics, 2022)
17. AI-Assisted Judicial Decision-Making: AI algorithms can aid judges in analyzing legal precedents and case data, supporting more informed and fair judicial decisions. (Reference: "AI in Judicial Decision-Making," American Bar Association, 2021)
18. AI-Driven Regulatory Compliance: AI can assist regulatory agencies in monitoring compliance with safety and quality standards in various sectors.

(Reference: "AI for Regulatory Compliance," European Union Agency for Cybersecurity, 2020)
19. AI-Powered Urban Planning: AI can analyze data on urban infrastructure and demographics to inform sustainable city planning and development. (Reference: "AI in Urban Planning," Smart Cities Council, 2022)
20. AI-Enhanced Human Resource Management: AI algorithms can improve talent acquisition, employee engagement, and workforce planning in government agencies. (Reference: "AI in Human Resources Management," Deloitte, 2021)
21. AI for Wildlife Conservation: AI-powered camera traps and image recognition can aid in monitoring and protecting endangered species. (Reference: "AI in Wildlife Conservation," World Wildlife Fund, 2019)
22. AI-Assisted Immigration Services: AI chatbots can provide personalized guidance and support to immigrants navigating the immigration process. (Reference: "AI for Immigration Services," International Organization for Migration, 2020)
23. AI-Powered Procurement: AI can optimize government procurement processes, ensuring cost-effectiveness and transparency. (Reference: "AI in Public Procurement," World Bank, 2021)
24. AI-Driven Public Health Surveillance: AI algorithms can analyze health data to detect disease outbreaks and monitor public health trends. (Reference: "AI in Public Health Surveillance," World Health Organization, 2022)
25. AI for Cultural Heritage Preservation: AI can aid in restoring and preserving cultural heritage artifacts

through digital reconstruction. (Reference: "AI for Cultural Heritage Preservation," UNESCO, 2021)
26. AI-Enabled Fraud Detection: AI algorithms can identify fraudulent activities in government programs, protecting public funds. (Reference: "AI for Fraud Detection," U.S. Government Accountability Office, 2021)
27. AI-Enhanced Transportation Planning: AI can optimize transportation networks and services, reducing congestion and enhancing mobility. (Reference: "AI in Transportation Planning," International Transport Forum, 2020)
28. AI for Water Resource Management: AI-powered sensors and data analytics can optimize water usage and monitor water quality. (Reference: "AI in Water Resource Management," United Nations Economic Commission for Europe, 2021)
29. AI-Powered Disaster Preparedness: AI can analyze historical disaster data to improve preparedness and response strategies. (Reference: "AI for Disaster Preparedness," Federal Emergency Management Agency, 2022)
30. AI-Driven Weather Forecasting: AI algorithms can enhance weather prediction accuracy, improving early warning systems for extreme weather events. (Reference: "AI in Weather Forecasting," World Meteorological Organization, 2021)

These few specific use cases offer a glimpse into the boundless potential of AI in transforming government services and public governance. As we move towards the future, integrating AI in the public sector promises to create more efficient, responsive, and citizen-centric government systems. AI-powered virtual assistants seamlessly assisting citizens with queries and tasks, reminiscent of science fiction visions where citizens interact with intelligent digital entities effortlessly.

AI-driven platforms provide personalized information and services, fostering a sense of empowerment and inclusion for all citizens. Moreover, the fictional references depict governments harnessing AI's data analytics capabilities to gain valuable insights into citizens' needs and preferences. Imagine governments using AI to analyze vast amounts of data, predicting and addressing societal challenges in a way that was once considered purely futuristic. From optimizing public transportation routes to predicting public health outbreaks and natural disasters.

AI-driven predictive analytics empowers governments to take proactive measures and make informed decisions. These fictional scenarios highlight a future where AI collaborates with human governance, making public services more tailored, efficient, and responsive to citizens' evolving needs. As the potential of AI in the public sector becomes increasingly apparent, the integration of AI in government services holds the promise of unlocking unprecedented levels of effectiveness and citizen satisfaction.

The Promise of AI

Mavericks' AI leaders

AI Leaders and Their Vision for Humanity's Future

Artificial Intelligence (AI) has captured the imagination of the world's brightest minds, propelling us into a future. Where technology and humanity merge in unprecedented ways. Let's take a glimpse into the visions of some of AI leaders and their profound perspectives on how AI will impact humanity.

1. Sundar Pichai (CEO of Google and Alphabet Inc.): Sundar Pichai envisions AI as a powerful force for positive change. He believes that AI's potential lies in addressing global challenges, from healthcare to environmental conservation. By harnessing AI responsibly, Pichai sees the opportunity to create a more inclusive and equitable world where technology empowers individuals and fosters collective progress.
2. Elon Musk (CEO of SpaceX and Tesla): Elon Musk acknowledges AI's immense potential but also warns about its risks. He emphasizes the need for

careful regulation and ethical considerations in AI development. Musk believes that unbridled AI advancement without proper safeguards could pose significant threats to humanity's future. However, he remains optimistic that, with responsible governance, AI can enhance our lives and propel us to new frontiers.
3. Fei-Fei Li (AI Researcher and Professor at Stanford University): Fei-Fei Li emphasizes the importance of AI being "inclusive, diverse, and representative" of humanity. She envisions AI as a tool that amplifies human capabilities rather than replacing them. Li believes that AI should be rooted in empathy, understanding, and the ethical integration of diverse perspectives to ensure its benefits reach all corners of society.
4. Satya Nadella (CEO of Microsoft): Satya Nadella sees AI as a transformative force that can drive productivity and elevate human potential. He advocates for responsible AI that empowers individuals, preserves privacy, and fosters trust. Nadella believes that AI's greatest impact will be in augmenting human capabilities, unleashing creativity, and solving challenges in areas like healthcare, education, and sustainability.
5. Demis Hassabis (CEO of DeepMind): Demis Hassabis envisions AI as a symbiotic extension of human capabilities. He believes that AI can unlock new discoveries in science, medicine, and beyond, working in tandem with human experts. Hassabis emphasizes the need for AI to be transparent, understandable, and accountable, paving the way for a future where AI-driven innovations enhance our collective understanding of the world.

6. Ginni Rometty (Former CEO of IBM): Ginni Rometty advocates for AI to be a tool for human progress rather than a replacement for human labor. She sees AI as a catalyst for creating new job opportunities, fostering innovation, and addressing societal challenges. Rometty believes that responsible AI development and collaboration between businesses, governments, and academia will be essential to ensure AI's benefits are accessible to all.
7. Sam Altman (CEO of OpenAI): Sam Altman is a visionary leader in the field of AI, shaping the future of AI research and its impact on humanity. Altman envisions AI as a force for positive change, enhancing human capabilities and addressing global challenges. He emphasizes the importance of aligning AI development with human values and ensuring that AI technologies benefit all of humanity. Altman's dedication to ensuring transparent and accountable AI research at OpenAI exemplifies his commitment to fostering responsible AI innovations that benefit society as a whole.
8. Andrew Ng (Founder of deeplearning.ai): Andrew Ng is a prominent figure in AI research and education, aiming to make AI accessible to all. As the founder of deeplearning.ai, he believes that democratizing AI knowledge is crucial to empower individuals and organizations across different fields. Ng emphasizes the importance of AI education to foster a broad understanding of AI's capabilities and limitations, enabling informed decision-making and responsible AI adoption. His dedication to AI education has had a transformative impact on

aspiring AI professionals and enthusiasts worldwide, unlocking new possibilities for innovation and societal progress.

AI leaders have diverse and insightful perspectives on the future of AI and its impact on humanity. While acknowledging AI's transformative potential, they also stress the importance of responsible development, ethics, and inclusivity. By harnessing AI for the collective good and maintaining a human-centric approach, we can pave the way for a future where AI and humanity thrive hand in hand, shaping a world that holds untold possibilities for progress and prosperity.

Resources

As you conclude this AI adventure, the Appendix provides a treasure trove of valuable resources and further reading materials to deepen your understanding of Artificial Intelligence. Whether you're a curious learner, an aspiring AI enthusiast, or a seasoned professional, these resources will equip you with additional knowledge and insights to continue your exploration of the AI frontier.

Books:

- "Artificial Intelligence: A Guide for Thinking Humans" by Melanie Mitchell
- "Human Compatible: Artificial Intelligence and the Problem of Control" by Stuart Russell
- "The Hundred-Page Machine Learning Book" by Andriy Burkov
- "AI Superpowers: China, Silicon Valley, and the New World Order" by Kai-Fu Lee

Online Courses and Tutorials:
- Coursera's "Machine Learning" by Andrew Ng
- edX's "AI for Everyone" by Andrew Ng
- Fast.ai's Practical Deep Learning for Coders
- TensorFlow's Tutorials and Guides

Research Papers and Journals:
- arXiv.org: A preprint repository with cutting-edge AI research papers from top academics and researchers.
- Nature: A reputable journal featuring AI-related articles and research breakthroughs.

AI Organizations and Initiatives:
- OpenAI: A research organization focused on promoting friendly AI that benefits humanity.
- DeepMind: A leading AI research lab dedicated to solving real-world problems.
- Partnership on AI: A multi-stakeholder initiative working to ensure AI benefits all of humanity.

AI Conferences and Events:
- NeurIPS (Conference on Neural Information Processing Systems)
- AAAI (Association for the Advancement of Artificial Intelligence) Conference
- CVPR (Computer Vision and Pattern Recognition) Conference

Online AI Communities:
- Reddit's /Machine Learning: A vibrant community discussing AI and machine learning topics.
- AI Stack Exchange: A Q&A platform where AI enthusiasts and experts exchange knowledge.

AI Software and Frameworks:
- TensorFlow: An open-source machine learning framework developed by Google.
- PyTorch: A popular deep learning framework supported by Facebook AI Research.
- scikit-learn: A versatile machine learning library for Python.
- AI News and Magazines:
- MIT Technology Review: Features AI news, insights, and thought-provoking articles.
- AI News: A platform offering the latest updates and trends in AI and machine learning.

Ethical AI Guidelines:
- The Ethics Guidelines for Trustworthy AI by the European Commission.
- The Asilomar AI Principles by Future of Life Institute.

These resources are a gateway to an ever-evolving world of AI knowledge and advancements. Dive in, explore, and stay curious as you continue your AI journey. Remember, the collective contributions of curious minds like yours shape the future of AI. Let this Appendix be your compass as you embark on new adventures in Artificial Intelligence.

The Promise of AI

AI-Powered Tools for policymakers

Integrating AI-powered software tools has become an indispensable catalyst for progress in the rapidly evolving landscape of government services and administration. These innovative tools are revolutionizing federal and local agencies' operations, empowering them to leverage artificial intelligence for various applications. From optimizing public service delivery to enhancing decision-making processes, AI is shaping the future of governance and citizen engagement. In this comprehensive exploration, we delve into a curated collection of AI-powered software tools that hold the potential to transform the government sector. Each tool offers unique capabilities and applications, providing government agencies with the means to streamline operations, harness the power of data, and improve overall efficiency. As we navigate through an array of AI-driven solutions, unveiling their transformative impact on government services and citizen experiences.

Here are more AI-powered software tools with detailed descriptions of their applications in the government sector, both at the federal and local levels:

1. Tabnine -
   Website: https://www.tabnine.com/ Description: Tabnine is an AI-powered code completion tool that offers intelligent suggestions as developers write code, enhancing productivity and accuracy. Relevance to Government: Tabnine can be invaluable for government developers and programmers, streamlining code development and improving software efficiency.
2. Copilot -
   Website: https://copilot.github.com/ Description: Copilot, developed by GitHub and OpenAI, is an advanced AI code generator that assists developers by suggesting whole lines of code in real-time. Relevance to Government: Copilot can significantly boost code development speed and quality for government software projects, improving overall efficiency.
3. Sourcery -
   Website: https://sourcery.ai/ Description: Sourcery is an AI-powered code refactoring tool that automatically detects and suggests code optimizations to enhance maintainability and readability. Relevance to Government: Sourcery can help government software teams maintain high code standards and streamline software maintenance processes.
4. Chat GPT - Website: N/A (OpenAI GPT-3 API) Description: Chat GPT is a conversational AI model developed by OpenAI, capable of engaging

in human-like dialogues and responding to user queries. Relevance to Government: Chat GPT can be utilized by government entities to create interactive chatbots for citizen engagement and support services.
5. Claude2 - Website: N/A (OpenAI GPT-3 API) Description: Claude2 is another AI model based on OpenAI's GPT-3, enabling natural language processing tasks like language translation and summarization. Relevance to Government: Claude2 can be applied to language-related government tasks, such as document translation and summarizing legal texts.
6. 11labs - Website: https://www.11labs.ai/ Description: 11labs provides AI-driven speech-to-text solutions, converting audio data into written text with high accuracy. Relevance to Government: Government agencies can utilize 11labs to transcribe audio records from meetings, interviews, and public hearings for documentation.
7. Rasa - Website: https://rasa.com/ Description: Rasa is a commercially supported platform for developing AI-powered conversational chatbots and virtual assistants. Relevance to Government: Rasa can empower government agencies to deploy sophisticated virtual assistants for citizen interactions and customer support.
8. Lang Chain - Website: N/A Description: Lang Chain is an AI platform that facilitates multilingual communication by providing real-time language translation services. Relevance to Government: Lang Chain can aid government entities in

overcoming language barriers during international interactions and diplomatic engagements.
9. Awesome Lang - Website: https://github.com/kevinlu1248/awesome-lang Description: Awesome Lang is a curated list of AI language processing libraries, tools, and resources for various natural language tasks. Relevance to Government: Government developers can explore Awesome Lang for AI language processing resources to enhance public communication and document analysis.
10. Awesome Fast API - Website: https://github.com/mjhea0/awesome-fastapi Description: Awesome Fast API is a collection of resources and tools for FastAPI, a modern Python web framework for building APIs quickly and efficiently. Relevance to Government: FastAPI can be employed by government agencies to develop robust and scalable APIs for their web services and applications.
11. Awesome GPT - Website: https://github.com/elyase/awesome-gpt Description: Awesome GPT is a curated list of resources and applications related to GPT models and AI language generation. Relevance to Government: Government researchers and AI practitioners can explore Awesome GPT for insights and ideas in deploying GPT-based solutions.
12. Dot GPT - Website: N/A (OpenAI GPT-3 API) Description: Dot GPT is an AI model based on GPT-3, specializing in natural language generation for various applications. Relevance to Government: Dot GPT can assist government entities in

generating text for reports, policy documents, and public communication.

13. Ecoute - Website: N/A Description: Ecoute is an AI-powered audio transcription tool that converts spoken content into written text, supporting multiple languages. Relevance to Government: Ecoute can be employed by government agencies to transcribe audio records, such as court proceedings and legislative sessions.

14. Knowledge GPT - Website: N/A (OpenAI GPT-3 API) Description: Knowledge GPT is an AI model capable of answering complex questions and providing detailed explanations on a wide range of topics. Relevance to Government: Knowledge GPT can be integrated into government websites and platforms to provide instant answers to citizen inquiries.

15. GPTQ - Website: N/A (OpenAI GPT-3 API) Description: GPTQ is an AI model based on GPT-3, specializing in generating questions and quizzes for educational and informational purposes. Relevance to Government: GPTQ can be utilized by government agencies for creating interactive quizzes and educational content for citizen engagement.

16. QLORA
Website: https://qlora.com/ Description: QLORA is an AI-driven analytics platform that assists in generating business insights from data. Relevance to Government: Government entities can utilize QLORA to gain valuable insights from large datasets, optimizing decision-making processes.

17. Falcon 2 - Website: https://www.tryfalcon.com/ Description : Falcon 2 is an AI-driven data visualization tool that simplifies complex datasets into interactive visual representations. Relevance to Government: Government analysts and policymakers can use Falcon 2 to present data in a more accessible and informative manner.
18. Test Generation UI - Website: N/A Description: Test Generation UI is an AI-driven tool that automates software testing and generates test cases for code analysis. Relevance to Government: Government software developers can benefit from Test Generation UI to ensure robust and error-free software applications.
19. Cogenosys - Website: https://www.cogenosys.com/ Description: Cogenosys is an AI-powered platform that automates data collection and analysis, providing valuable insights for business decisions. Relevance to Government: Cogenosys can be used by government agencies for data-driven policy development and performance monitoring.
20. AutoGPT - Website: N/A (OpenAI GPT-3 API) Description: AutoGPT is an AI model specialized in automating repetitive tasks and generating content based on specific user inputs. Relevance to Government: AutoGPT can streamline government content generation, such as form responses, FAQs, and informational documents.
21. ThinkGPT - Website: N/A (OpenAI GPT-3 API) Description: ThinkGPT is an AI model that focuses on brainstorming ideas and generating creative concepts based on user input. Relevance to

Government: ThinkGPT can assist government entities in idea generation for innovative projects and problem-solving initiatives.
22. ChromaGPT - Website: N/A (OpenAI GPT-3 API) Description: ChromaGPT is an AI model designed to generate music and harmonies, making it useful for creative content creation. Relevance to Government: ChromaGPT can be employed by government agencies to produce original soundtracks for promotional materials or public campaigns.
23. Godmode Space - Website: https://godmode.space/ Description: Godmode Space is an AI-driven platform for automated debugging and troubleshooting in software development. Relevance to Government: Government software teams can utilize Godmode Space to expedite the debugging process and improve software reliability.
24. AudioGPT - Website: N/A (OpenAI GPT-3 API) Description: AudioGPT is an AI model focused on generating audio content, enabling applications like voice synthesis and music composition. Relevance to Government: AudioGPT can be utilized by government agencies for creating audio content for public service announcements and educational materials.
25. ChatPDF - Website: N/A (OpenAI GPT-3 API) Description: ChatPDF is an AI model that interacts with PDF documents, providing information and answering questions based on document content. Relevance to Government: ChatPDF can assist government agencies in extracting relevant

information from lengthy documents and facilitating document accessibility.
26. GitHub Awesome AI List - Website: https://github.com/owainlewis/awesome-artificial-intelligence Description: GitHub Awesome AI List is a curated compilation of AI resources, frameworks, and tools for developers and researchers. Relevance to Government: Government developers and AI researchers can explore this list for a wide range of AI-related tools and libraries.
27. GitHub Lang Chain List - Website: https://github.com/kevinlu1248/lang-chain Description: GitHub Lang Chain List is a collection of AI language processing tools and resources for natural language understanding and generation. Relevance to Government: Government AI researchers can find valuable language processing resources in this list for text analysis and AI-driven language applications.
28. DevGPT - Website: N/A (OpenAI GPT-3 API) Description: DevGPT is an AI model specialized in assisting developers with code-related tasks, such as code generation and debugging support. Relevance to Government: DevGPT can be employed by government developers to streamline code development and improve software reliability.
29. Deep Lake - Website: https://deeplake.io/ Description: Deep Lake offers an AI platform for image and video analysis, enabling applications like object recognition and content moderation. Relevance to Government: Government agencies can leverage

Deep Lake for image analysis tasks in surveillance, disaster response, and public safety.
30. Quivr Codex - Website: N/A (OpenAI GPT-3 API) Description: Quivr Codex is an AI model that generates detailed explanations for complex concepts and technical documentation. Relevance to Government: Quivr Codex can assist government entities in producing user-friendly technical documents and policy explanations.
31. Civicompass - Website: https://civicompass.com/ Description: Civicompass is an AI-powered platform that offers automated sentiment analysis and opinion mining for public opinion insights. Relevance to Government: Civicompass can be used by government agencies to analyze public sentiment towards policies and programs.
32. PerceptiLabs - Website: https://www.perceptilabs.com/ Description: PerceptiLabs is an AI platform that simplifies the creation and visualization of deep learning models. Relevance to Government: Government researchers and data scientists can utilize PerceptiLabs for AI model development in various applications.
33. Govchat - Website: https://www.govchat.org/ Description: Govchat is an AI-powered chatbot platform tailored for government agencies, enabling personalized interactions with citizens. Relevance to Government: Govchat can improve citizen engagement and provide quick and accurate responses to frequently asked questions.
34. DRYiCE Lucy - Website: https://www.dryice.ai/ Description: DRYiCE Lucy is an AI-powered virtual assistant

designed to support IT service management and automate routine tasks. Relevance to Government: Government IT departments can use DRYiCE Lucy to enhance IT support services and optimize IT operations.

35. Zindi - Website: https://zindi.africa/ Description: Zindi is an AI competition platform that hosts data science challenges to solve real-world problems. Relevance to Government: Government entities can engage with Zindi to crowdsource innovative AI solutions for complex societal issues.

36. Babbel - Website: https://www.babbel.com/ Description: Babel is an AI-powered language learning platform that assists users in mastering new languages. Relevance to Government: Government agencies can use Babel for language training and communication support in international relations.

37. OpenDataSoft - Website: https://www.opendatasoft.com/ Description: OpenDataSoft is an AI platform that enables governments to manage and share open data with the public. Relevance to Government: OpenDataSoft helps government agencies promote transparency and engage citizens with accessible data.

38. Granulate - Website: https://www.granulate.io/ Description: Granulate offers an AI-driven platform for optimizing application performance and enhancing user experiences. Relevance to Government: Government websites and applications can benefit from Granulate's performance optimization to ensure seamless user experiences.

39. DeepDive
Website: https://deepdive.ai/ Description: DeepDive is an AI platform for advanced data analytics and predictive modeling. Relevance to Government: Government agencies can use DeepDive for data-driven decision-making and predicting trends for various sectors.
40. Suki.AI - Website: https://suki.ai/ Description: Suki.AI is an AI-powered medical assistant that provides real-time clinical documentation support to healthcare professionals. Relevance to Government: Suki.AI can be integrated into government healthcare systems to streamline clinical documentation and improve patient care.
41. Insight Engines
Website: https://www.insightengines.com/ Description: Insight Engines is an AI platform that provides natural language search capabilities for enterprise data exploration. Relevance to Government: Government agencies can use Insight Engines to analyze vast datasets and extract relevant information for decision-making.
42. Emotibot
Website: https://www.emotibot.com/ Description: Emotibot is an AI-powered virtual assistant that specializes in emotion recognition and sentiment analysis. Relevance to Government: Emotibot can enhance government virtual assistants and chatbots, enabling more personalized interactions with citizens.
43. ThryveAI
Website: https://www.thryve.ai/ Description: ThryveAI is an AI-driven platform for personalized nutrition and health recommendations based on

individual data. Relevance to Government: Government healthcare initiatives can leverage ThryveAI to promote personalized health recommendations and preventive measures.
44. Viz.ai - Website: https://www.viz.ai/ Description: Viz.ai is an AI platform for analyzing medical images, with a focus on time-critical conditions like stroke detection. Relevance to Government: Viz.ai can support government healthcare facilities in faster and more accurate diagnosis of critical medical conditions.

As the AI landscape evolves, these AI applications and platforms can significantly impact government entities, fostering innovation, enhancing public services, and driving informed decision-making. Government agencies need to explore and integrate AI tools that align with their specific needs and objectives, ensuring the responsible and ethical application of AI technologies in serving the public interest.

Assess Your AI Skills

1. What is the term commonly used to describe periods of decreased funding and interest in AI research?
    a. AI Renaissance
    b. AI Enlightenment
    c. AI Spring
    d. AI Winter
2. Who is often referred to as the "Father of Artificial Intelligence" for his groundbreaking work in computer science and AI research?
    a. Alan Turing
    b. John McCarthy
    c. Marvin Minsky
    d. Herbert Simon
3. Which AI technology is used to understand and analyze natural language input from humans and provide relevant responses?
    a. Neural Networks
    b. Genetic Algorithms
    c. Expert Systems
    d. Natural Language Processing (NLP)

4. What is the purpose of bias mitigation in AI systems?
   a. To increase bias in AI decision-making
   b. To ensure AI systems are free from any form of bias
   c. To amplify existing biases in AI models
   d. To balance bias between different demographic groups
5. Which real-world case highlighted the ethical concern of AI systems discriminating against female candidates during the hiring process?
   a. Amazon's Gender-Biased Hiring Tool
   b. ProPublica's Investigation into COMPAS Algorithm
   c. European Union's General Data Protection Regulation (GDPR)
   d. Google's Project Magenta
6. What is the main objective of the European Union's General Data Protection Regulation (GDPR)?
   a. To promote data sharing without any restrictions
   b. To empower citizens with control over their personal data
   c. To encourage organizations to collect more user data
   d. To minimize AI development in the EU
7. Which visionary AI leader envisions AI as a symbiotic extension of human capabilities and emphasizes transparency and accountability?
   a. Ginni Rometty
   b. Demis Hassabis
   c. Elon Musk
   d. Andrew Ng

8. What is the key objective of the Organization for Economic Cooperation and Development (OECD) AI Principles?
    a. To promote AI monopolies in the industry
    b. To establish universal AI regulations
    c. To provide a common framework for responsible AI policy
    d. To ban the use of AI in government operations
9. How does the concept of "flat world" relate to the international collaboration in AI governance?
    a. It emphasizes the need for AI to be complex and inaccessible.
    b. It signifies the global interconnectedness that requires collaborative AI governance.
    c. It suggests that AI should only be governed by individual countries.
    d. It promotes isolationism in AI development.
10. Which AI technology focuses on generating creative content like music composition and strives for diversity in training data?
    a. Natural Language Processing (NLP)
    b. Reinforcement Learning
    c. Genetic Algorithms
    d. Google's Project Magenta
11. What is the main purpose of the "AI Adventures" book?
    a. To explore the history of AI research
    b. To provide guidelines for policymakers in AI adoption
    c. To showcase AI's impact on government and society

    d. To highlight the challenges of AI ethics
12. Which AI leader envisions AI as a tool for human progress, creating new job opportunities and fostering innovation?
    a. Elon Musk
    b. Demis Hassabis
    c. Ginni Rometty
    d. Andrew Ng
13. What is the focus of Part II in the book "Unveiling the AI Horizon"?
    a. Exploring AI's promise for governments and citizens
    b. Understanding AI's history and evolution
    c. Investigating AI ethics and bias mitigation
    d. Creating an AI framework for policymakers
14. What is the key objective of the Draft National Deep Tech Startup Policy?
    a. To discourage deep tech startups in India
    b. To attract global investors to Indian startups
    c. To promote AI technologies exclusively
    d. To nurture India's deep tech startup ecosystem
15. Who is the GoI Principal Scientific Advisor associated with the Draft National Deep Tech Startup Policy?
    a. Demis Hassabis
    b. Ajay Sood
    c. Ginni Rometty
    d. Andrew Ng
16. In AI policymaking, what does "IP" stand for?
    a. Intelligent Programming
    b. International Policy

c. Intellectual Property
d. Intelligent Policy
17. What is the main goal of the "AI Adventures" book?
    a. To dive into the history of AI research
    b. To guide businesses in AI adoption
    c. To explore the challenges of AI ethics
    d. To provide a roadmap for AI governance
18. What are the ethical considerations when adopting AI in policymaking?
    a. Ignoring privacy concerns
    b. Overlooking bias and transparency
    c. Focusing solely on financial benefits
    d. Prioritizing AI development over human rights
19. How can global policy cooperation impact AI governance in a "flat world"?
    a. By isolating AI development within individual countries
    b. By fostering collaboration in AI regulation and ethics
    c. By centralizing AI development in a single nation
    d. By restricting AI technologies to certain industries
20. What is the primary focus of AI in education?
    a. Enhancing customer interactions
    b. Promoting financial optimization
    c. Fostering sustainable development
    d. Enriching the learning experience
21. Which AI leader envisions AI as an extension of human capabilities and emphasizes transparency and accountability?
    a. Demis Hassabis

b. Ginni Rometty
c. Elon Musk
d. Andrew Ng
22. What is the main purpose of bias mitigation in AI systems?
    a. To amplify existing biases
    b. To balance bias between demographic groups
    c. To increase bias in decision-making
    d. To ensure fairness and impartiality
23. Which real-world case highlighted the bias in AI decision-making during the hiring process?
    a. Amazon's Gender-Biased Hiring Tool
    b. ProPublica's Investigation into COMPAS Algorithm
    c. European Union's General Data Protection Regulation (GDPR)
    d. Google's Project Magenta
24. What is the central objective of the European Union's General Data Protection Regulation (GDPR)?
    a. To encourage organizations to collect more user data.
    b. To promote unrestricted data sharing
    c. To empower citizens with control over personal data
    d. To minimize AI development in the EU
25. How does the "flat world" concept relate to international collaboration in AI governance?
    a. It emphasizes isolationism in AI development
    b. It signifies the global interconnectedness for collaborative AI governance.

c. It promotes exclusive AI development within individual countries.
d. It suggests that AI governance is unnecessary on a global scale.

26. What is the primary goal of the Organization for Economic Cooperation and Development (OECD) AI Principles?
    a. To establish universal AI regulations
    b. To promote AI monopolies in the industry
    c. To provide a common framework for responsible AI policy
    d. To ban the use of AI in government operations

27. Which visionary AI leader envisions AI as a symbiotic extension of human capabilities?
    a. Demis Hassabis
    b. Ginni Rometty
    c. Elon Musk
    d. Andrew Ng

28. What is the central theme of Part III in the book "Unveiling the AI Horizon"?
    a. Understanding the history of AI research
    b. Exploring AI ethics and bias mitigation
    c. Investigating AI's impact on governments and citizens
    d. Crafting an AI framework for policymakers

29. What is the focus of the "Draft National Deep Tech Startup Policy"?
    a. Encouraging global investors to Indian startups
    b. Discouraging deep tech startups in India
    c. Promoting AI technologies exclusively

    d. Nurturing India's deep tech startup ecosystem
30. Who is associated with the "Draft National Deep Tech Startup Policy" as its spearhead?
    a. Demis Hassabis
    b. Ajay Sood
    c. Ginni Rometty
    d. Andrew Ng
31. In AI policymaking, what does "IP" stand for?
    a. Intelligent Programming
    b. International Policy
    c. Intellectual Property
    d. Intelligent Policy
32. What is the primary aim of the "AI Adventures" book?
    a. To explore AI ethics and bias mitigation
    b. To guide businesses in AI adoption
    c. To delve into the history of AI research
    d. To provide a roadmap for AI governance
33. What are some of the ethical considerations when adopting AI in policymaking?
    a. Ignoring privacy concerns
    b. Overlooking bias and transparency
    c. Focusing solely on financial benefits
    d. Prioritizing AI development over human rights
34. How can global policy cooperation impact AI governance in a "flat world"?
    a. By centralizing AI development in a single nation
    b. By restricting AI technologies to certain industries
    c. By fostering collaboration in AI regulation and ethics

d. By isolating AI development within individual countries
35. What is the primary focus of AI in education?
    a. Enhancing customer interactions
    b. Promoting financial optimization
    c. Fostering sustainable development
    d. Enriching the learning experience
36. Which AI leader envisions AI as an extension of human capabilities and emphasizes transparency and accountability?
    a. Demis Hassabis
    b. Ginni Rometty
    c. Elon Musk
    d. Andrew Ng
37. What is the main purpose of bias mitigation in AI systems?
    a. To amplify existing biases
    b. To balance bias between demographic groups
    c. To increase bias in decision-making
    d. To ensure fairness and impartiality
38. Which real-world case highlighted the bias in AI decision-making during the hiring process?
    a. Amazon's Gender-Biased Hiring Tool
    b. ProPublica's Investigation into COMPAS Algorithm
    c. European Union's General Data Protection Regulation (GDPR)
    d. Google's Project Magenta
39. What is the central objective of the European Union's General Data Protection Regulation (GDPR)?
    a. To encourage organizations to collect more user data.

b. To promote unrestricted data sharing
c. To empower citizens with control over personal data
d. To minimize AI development in the EU

40. How does the "flat world" concept relate to international collaboration in AI governance?
    a. It emphasizes isolationism in AI development.
    b. It signifies the global interconnectedness for collaborative AI governance.
    c. It promotes exclusive AI development within individual countries.
    d. It suggests that AI governance is unnecessary on a global scale.

41. What is the primary goal of the Organization for Economic Cooperation and Development (OECD) AI Principles?
    a. To establish universal AI regulations
    b. To promote AI monopolies in the industry
    c. To provide a common framework for responsible AI policy
    d. To ban the use of AI in government operations

42. Which visionary AI leader envisions AI as a symbiotic extension of human capabilities?
    a. Demis Hassabis
    b. Ginni Rometty
    c. Elon Musk
    d. Andrew Ng

43. What is the central theme of Part III in the book "Unveiling the AI Horizon"?
    a. Understanding the history of AI research
    b. Exploring AI ethics and bias mitigation

c. Investigating AI's impact on governments and citizens
d. Crafting an AI framework for policymakers

44. What is the focus of the "Draft National Deep Tech Startup Policy"?
    a. Encouraging global investors to Indian startups
    b. Discouraging deep tech startups in India
    c. Promoting AI technologies exclusively
    d. Nurturing India's deep tech startup ecosystem

45. Who is associated with the "Draft National Deep Tech Startup Policy" as its spearhead?
    a. Demis Hassabis
    b. Ajay Sood
    c. Ginni Rometty
    d. Andrew Ng

46. In AI policymaking, what does "IP" stand for?
    a. Intelligent Programming
    b. International Policy
    c. Intellectual Property
    d. Intelligent Policy

47. What is the primary aim of the "AI Adventures" book?
    a. To explore AI ethics and bias mitigation
    b. To guide businesses in AI adoption
    c. To delve into the history of AI research
    d. To provide a roadmap for AI governance

48. What are some of the ethical considerations when adopting AI in policymaking?
    a. Ignoring privacy concerns
    b. Overlooking bias and transparency
    c. Focusing solely on financial benefits

d. Prioritizing AI development over human rights
49. How can global policy cooperation impact AI governance in a "flat world"?
    a. By centralizing AI development in a single nation
    b. By restricting AI technologies to certain industries
    c. By fostering collaboration in AI regulation and ethics
    d. By isolating AI development within individual countries
50. What is the primary focus of AI in education?
    a. Enhancing customer interactions
    b. Promoting financial optimization
    c. Fostering sustainable development
    d. Enriching the learning experience
51. Which AI leader envisions AI as an extension of human capabilities and emphasizes transparency and accountability?
    a. Demis Hassabis
    b. Ginni Rometty
    c. Elon Musk
    d. Andrew Ng
52. What is the main purpose of bias mitigation in AI systems?
    a. To amplify existing biases
    b. To balance bias between demographic groups
    c. To increase bias in decision-making
    d. To ensure fairness and impartiality.
53. Which real-world case highlighted the bias in AI decision-making during the hiring process?
    a. Amazon's Gender-Biased Hiring Tool

b. ProPublica's Investigation into COMPAS Algorithm
   c. European Union's General Data Protection Regulation (GDPR)
   d. Google's Project Magenta
54. What is the central objective of the European Union's General Data Protection Regulation (GDPR)?
   a. To encourage organizations to collect more user data.
   b. To promote unrestricted data sharing
   c. To empower citizens with control over personal data
   d. To minimize AI development in the EU
55. How does the "flat world" concept relate to international collaboration in AI governance?
   a. It emphasizes isolationism in AI development.
   b. It signifies the global interconnectedness for collaborative AI governance.
   c. It promotes exclusive AI development within individual countries.
   d. It suggests that AI governance is unnecessary on a global scale.
56. What is the primary goal of the Organization for Economic Cooperation and Development (OECD) AI Principles?
   a. To establish universal AI regulations
   b. To promote AI monopolies in the industry
   c. To provide a common framework for responsible AI policy
   d. To ban the use of AI in government operations.

57. Which visionary AI leader envisions AI as a symbiotic extension of human capabilities?
    a. Demis Hassabis
    b. Ginni Rometty
    c. Elon Musk
    d. Andrew Ng
58. What is the central theme of Part III in the book "Unveiling the AI Horizon"?
    a. Understanding the history of AI research
    b. Exploring AI ethics and bias mitigation
    c. Investigating AI's impact on governments and citizens
    d. Crafting an AI framework for policymakers
59. What is the focus of the "Draft National Deep Tech Startup Policy"?
    a. Encouraging global investors to Indian startups
    b. Discouraging deep tech startups in India
    c. Promoting AI technologies exclusively
    d. Nurturing India's deep tech startup ecosystem
60. Who is associated with the "Draft National Deep Tech Startup Policy" as its spearhead?
    a. Demis Hassabis
    b. Ajay Sood
    c. Ginni Rometty
    d. Andrew Ng
61. In AI policymaking, what does "IP" stand for?
    a. Intelligent Programming
    b. International Policy
    c. Intellectual Property
    d. Intelligent Policy

62. What is the primary aim of the "AI Adventures" book?
    a. To explore AI ethics and bias mitigation
    b. To guide businesses in AI adoption
    c. To delve into the history of AI research
    d. To provide a roadmap for AI governance
63. What are some of the ethical considerations when adopting AI in policymaking?
    a. Ignoring privacy concerns
    b. Overlooking bias and transparency
    c. Focusing solely on financial benefits
    d. Prioritizing AI development over human rights
64. How can global policy cooperation impact AI governance in a "flat world"?
    a. By centralizing AI development in a single nation
    b. By restricting AI technologies to certain industries
    c. By fostering collaboration in AI regulation and ethics
    d. By isolating AI development within individual countries
65. What is the primary focus of AI in education?
    a. Enhancing customer interactions
    b. Promoting financial optimization
    c. Fostering sustainable development
    d. Enriching the learning experience.
66. Which AI leader envisions AI as an extension of human capabilities and emphasizes transparency and accountability?
    a. Demis Hassabis
    b. Ginni Rometty
    c. Elon Musk

    d. Andrew Ng
67. What is the primary purpose of bias mitigation in AI systems?
    a. To amplify existing biases
    b. To balance bias between demographic groups
    c. To increase bias in decision-making
    d. To ensure fairness and impartiality
68. Which real-world case highlighted the bias in AI decision-making during the hiring process?
    a. Amazon's Gender-Biased Hiring Tool
    b. ProPublica's Investigation into COMPAS Algorithm
    c. European Union's General Data Protection Regulation (GDPR) d) Google's Project Magenta
69. What is the central objective of the European Union's General Data Protection Regulation (GDPR)?
    a. To encourage organizations to collect more user data.
    b. To promote unrestricted data sharing
    c. To empower citizens with control over personal data
    d. To minimize AI development in the EU
70. How does the "flat world" concept relate to international collaboration in AI governance?
    a. It emphasizes isolationism in AI development.
    b. It signifies the global interconnectedness for collaborative AI governance.
    c. It promotes exclusive AI development within individual countries.

The Promise of AI

      d. It suggests that AI governance is unnecessary on a global scale.

71. What is the primary goal of the Organization for Economic Cooperation and Development (OECD) AI Principles?
    a. To establish universal AI regulations
    b. To promote AI monopolies in the industry
    c. To provide a common framework for responsible AI policy
    d. To ban the use of AI in government operations.

72. Which visionary AI leader envisions AI as a symbiotic extension of human capabilities?
    a. Demis Hassabis
    b. Ginni Rometty
    c. Elon Musk
    d. Andrew Ng

73. What is the central theme of Part III in the book "Unveiling the AI Horizon"?
    a. Understanding the history of AI research
    b. Exploring AI ethics and bias mitigation
    c. Investigating AI's impact on governments and citizens
    d. Crafting an AI framework for policymakers

74. What is the focus of the "Draft National Deep Tech Startup Policy"?
    a. Encouraging global investors to Indian startups
    b. Discouraging deep tech startups in India
    c. Promoting AI technologies exclusively
    d. Nurturing India's deep tech startup ecosystem

75. Who is associated with the "Draft National Deep Tech Startup Policy" as its spearhead?
    a. Demis Hassabis
    b. Ajay Sood
    c. Ginni Rometty
    d. Andrew Ng

Answers

1. A
2. D
3. B
4. C
5. D
6. B
7. B
8. B
9. C
10. A
11. A
12. D
13. B
14. C
15. A
16. B
17. C
18. B
19. C
20. A
21. A
22. D
23. B
24. C
25. B
26. C
27. D
28. A
29. B
30. C
31. B
32. D
33. C
34. B
35. A
36. C
37. D
38. B
39. C
40. A
41. C
42. A
43. B
44. D
45. B

46. C	56. C	66. A
47. B	57. A	67. D
48. D	58. C	68. A
49. B	59. D	69. C
50. D	60. B	70. B
51. A	61. C	71. C
52. D	62. B	72. A
53. A	63. D	73. D
54. C	64. C	74. D
55. B	65. D	75. B

Level 1: Novice

0-25% correct answers

You're just getting started on your AI knowledge journey. Keep exploring and learning to unlock the potential of AI in various fields.

Level 2: Apprentice

26-50% correct answers

You have a basic understanding of AI concepts, but there's still plenty to discover. Dive deeper into AI applications and their impact.

Level 3: Enthusiast

51-75% correct answers

You're well-versed in AI fundamentals and its role in policy-making and governance. Keep up the great work, and continue expanding your AI expertise.

Level 4: Expert

76-100% correct answers

Congratulations! You have a strong grasp of AI concepts, their history, ethical considerations, and real-world applications. Your expertise can contribute significantly to shaping AI policies and strategies.

The Promise of AI

The Promise of AI

Acknowledgements

As I come to the end of adventure, I am filled with immense gratitude and joy for the support and inspiration that have fueled the creation of this book. With heartfelt appreciation, I extend my sincerest thanks to the remarkable individuals and organizations that have contributed to my journey. First and foremost, I am grateful to the visionary AI leaders and experts whose insights and innovative work have shaped the content of this book. I want to express my deep appreciation to the researchers, scientists, and engineers who tirelessly push the boundaries of AI, uncovering new possibilities that continue to amaze us all. Your pioneering work lays the foundation for a future that holds boundless promise.

To my friends and family, thank you for your unwavering support and encouragement throughout this writing process. Your belief in me and in the importance of sharing the AI journey has been an incredible source of motivation. A special thanks to my editor and the entire publishing team for their guidance and expertise in bringing this book to life. Your dedication to excellence and attention to detail have enriched the final product.

To the readers, thank you for joining me on this AI adventure. Your curiosity and open-mindedness are the driving force behind my desire to share the wonders of AI with the world. This book has ignited a spark of excitement and curiosity in your hearts. Lastly, I extend my deepest gratitude to the AI community. This vast network of passionate individuals and organizations is tirelessly working towards a future where AI empowers us to overcome challenges and create a better world.

To every person who has played a role, big or small, in this endeavor, I am profoundly grateful. Your support, encouragement, and contributions have made this book possible, and I am honored to have you by my side on this AI adventure.

With immense appreciation,

Sanjiv Goyal

Words for the readers

As I conclude this book, I am filled with an overwhelming sense of awe and excitement for the future that lies ahead. Our journey through the marvelous world of Artificial Intelligence has been nothing short of extraordinary, revealing the boundless potential that AI holds in shaping our world for the better.

Throughout these pages, we've witnessed AI in action, revolutionizing industries, enhancing lives, and safeguarding our environment. From autonomous Tesla vehicles navigating the roads with finesse to AI-driven virtual assistants anticipating our every need, we've explored a realm where imagination meets reality. But the true magic of AI lies not just in the incredible applications we've encountered but, in the opportunities, it offers each one of us. As we step away from these pages, I urge you to embrace the spirit of exploration and curiosity in your own AI adventures.

Whether you're a business leader seeking to unlock AI's potential in your organization, a student eager to dive into the world of AI research, or simply an individual curious about the possibilities that lie ahead, there is a place for you in this AI frontier. By harnessing AI's power responsibly and ethically, we can overcome challenges that once seemed insurmountable. Together, we can address global issues, foster inclusive growth, and build a more sustainable and equitable future.

As we look ahead, let's remember that the true adventure has just begun. The AI landscape is ever evolving, and the possibilities are limited only by our imagination. Embrace the unknown, for the most remarkable discoveries await within the realm of the unknown.

I extend my heartfelt gratitude to my fellow adventurers for joining me on this journey. This book has inspired you to envision the endless possibilities of AI and to take the first step toward your own AI adventure.

Let us not merely observe the unfolding AI revolution but actively shape it with our ingenuity and compassion. As we venture into the AI frontier, let us do so with a sense of responsibility, mindful of our impact on the world.

The future is calling, and with AI as our ally, we have the power to create a world that surpasses our wildest dreams. Embrace the AI adventure that awaits you, and together, let us shape a brighter and more extraordinary future for all.

Connect with me and follow my journey:

- www.linkedin.com/in/sanjivgoyal
- www.youtube.com/sanjivgoyal
- www.sanjivgoyal.com

Do share your thoughts, and let's explore insights, and vision for the future.

Warm regards,

Sanjiv Goyal

About the Author

Sanjiv Goyal is a visionary and thought leader in the realms of AI Entrepreneurship, Angel Investment, and futurism. Armed with an M Tech from the prestigious IIT Delhi, his journey embodies a deep-rooted passion for leveraging cutting-edge technology to catalyze positive societal change. His profound insights into the symbiotic relationship between AI and human potential position him as a pioneering figure in the AI landscape.

Sanjiv's vision transcends conventional boundaries, firmly anchored in the belief that AI is a potent tool to amplify human ingenuity and address pressing global challenges. His dedication to the ethical development of AI underscores the significance of transparency and accountability in shaping a future enriched by AI's transformative innovations.

As an influential board member and investor in trailblazing tech ventures, Sanjiv's advocacy for AI's pivotal role in societal advancement is resolute. He envisions AI not as a replacement for human endeavors, but as a catalyst for innovation and novel opportunities. Sanjiv's proactive approach empowers organizations to embrace AI responsibly, fostering societal progress.

Central to his ethos is the empowerment of diverse individuals through education. A mentor and supporter of educational initiatives, Sanjiv champions startups and AI journeys, ensuring accessibility and inclusivity. His commitment to human-centered AI underscores the alignment of technology with human values and needs.

Sanjiv's visionary foresight, unwavering ethical compass, and dedication to AI education drive a transformative paradigm shift. His ventures exemplify responsible AI innovation, inspiring the limitless boundaries of technology, and forging a future where human advancement flourishes in harmony with cutting-edge AI. His unyielding passion is a guiding light that illuminates the path towards a world transformed by the positive potential of technology.

Book Recommandations

The Author has explored the profound interplay of AI and governance in this thought-provoking odyssey. A must-read for everyone. - Najma Akhtar, VC, Jamia Millia Islamia, Delhi, India

It is a great foundational read on AI evolution and promise. Very insightful." - Sanjay Shringarpure, CIO, RNDC

This book showcases real-world examples of AI, from chatbots to sustainable transportation. Be inspired by how AI can revolutionize our world." – Mark Tonnesen, CIO, Achieve

The Author takes you on an enlightening exploration of AI's history and global implications. - Maya Strelar-Migotti, Angel Investor, Former EVP, Marvell Semiconductor

It is a guide that empowers businesses to harness AI for customer interactions, financial optimization, and innovation. Craft a tailored AI strategy and secure a competitive edge in a dynamic market. – Jacob Lanning, Innovation and Tech Leader, IGT.

This book provides relevant examples and a friendly tone to help you tailor AI solutions for your industry. Join the movement towards a future where AI and humanity thrive together. – CJ Sattler, CTO, LV.net

Explore its applications across finance, healthcare, education, and more. It is essential for those ready to embark on an exciting AI journey. - Amit Goel, CEO, Droisys, Inc.

The Author has presented a compelling framework for customer experiences, financial optimization, and innovation. – Rupam Shrivastava, Partner, Frontiers Fund.

Made in the USA
Middletown, DE
13 July 2024